CD-I and Interactive Videodisc Technology

CD-I and Interactive Videodisc Technology

Edited by
Steve Lambert and Jane Sallis

Howard W. Sams & Co.
A Division of Macmillan, Inc.
4300 West 62nd Street, Indianapolis, IN 46268 USA

FIRST EDITION
FIRST PRINTING—1986

International Standard Book Number: 0-672-22513-1
Library of Congress Catalog Card Number: 86-62443

Acquisitions Editor: *Greg Michael*
Designer: *T. R. Emrick*
Illustrator: *Don Clemons*
Cover Design: *Jerry Bates, Visual Graphic Services*
Cover Illustration: *Keith J. Hampton, Visual Graphic Services*
Compositor: *Impressions, Inc., Madison, Wisconsin*

Printed in the United States of America

Contents

Introduction

Our objective in compiling this book was to examine the similarities and differences between two of today's most exciting information media: interactive videodisc (IVD) and compact disc–interactive (CD-I). We chose to focus on IVD and CD-I because of their multisensory approach to interactivity, which incorporates the use of sound and motion video in addition to text and still-frame display.

The original premise of the book was that, though the technologies differ, they share the same objective: to deliver information and/or entertainment that is extremely responsive to the individual user. With this in mind, we structured the book to include chapters having a technical nature along with chapters delineating the user's role within the high-tech environment (or perhaps we should say the machine's role within the human environment).

All of our authors are recognized experts in their field, and, as such, their chapters reflect their personal experiences and points of view. We have tried to maintain the writing style of each author, and, because our intent was to make each chapter a discrete entity that could be read independently of any other, certain content redundancies have been retained. Our hope was that, upon concluding each chapter, the reader would feel as though he or she had had an actual conversation with that particular author.

Please bear in mind that the technical information contained in this book represents the state of the art as it existed in August of 1986. There is no reason to expect that the various limitations attributed to each technology within the context of this book will not be addressed by the next generation of videodisc and CD players—or by some new hybrid medium.

The rest of this introduction presents some background information on IVD. If you are already familiar with this type of information, feel free to proceed directly to the chapters of your choice.

INTERACTIVE VIDEODISCS—AN OVERVIEW

Laser videodisc technology as we know it today was developed almost simultaneously by MCA (Music Corporation of America) and NV Philips in the early

1970s. Also around this time, RCA developed SELECTAVISION®, a nonlaser capacitance electronic disc (CED) system which used a stylus that came in actual contact with the videodisc's surface. However, RCA abandoned this format in 1984, leaving laser videodiscs as the industry standard.

Information is encoded on a laser videodisc in the form of microscopic pits that have been pressed into the disc in a spiral configuration. (Each 360-degree segment of this spiral is called a *track*.) This information is then "read" by the laser beam inside the videodisc player and passed on to the monitor and/or microprocessor or computer.

Types of Laser Videodiscs

At present, there are two different types of laser videodiscs: reflective videodiscs and transmissive videodiscs.

Reflective Videodiscs

A *reflective videodisc* has an opaque, shiny surface. In order to read the material encoded on the disc, the laser beam is reflected off the surface of the disc onto a mirror that, in turn, reflects the beam into a decoder. The disc itself is covered with layers of protective plastic, which make it virtually impervious to damage.

Reflective videodiscs, under the trade name LaserVision, are the predominant videodisc format. A disc described as "nonstandard" is not compatible with the LaserVision format and cannot be played on LaserVision systems; conversely, a system described as nonstandard cannot play LaserVision discs.

Recent developments in reflective videodisc technology include DRAW (Direct Read After Write) discs, 8-inch discs, and disc players with "instant-jump" capabilities.

DRAW discs: These are reflective videodiscs that result from a real-time recording process that produces a single, immediately playable videodisc (as opposed to the standard pressing process, which requires that a master be made before replication can occur). DRAW discs are frequently called *check* discs because their principal use is for checking videodisc content prior to the more expensive standard replication process. The audio and video qualities of a DRAW disc are not up to the levels achieved through the standard replication process, but, then, they are not supposed to be. DRAW discs are meant to be part of the developmental process rather than the end product, and, in that respect, they have proved themselves invaluable.

At present, there are two DRAW disc systems on the market: Optical Disc Corporation's Recordable Laser Videodisc (RLV) and Panasonic's Optical Memory Disc Recorder (OMDR).

Optical Disc Corporation's RLV is compatible with LaserVision systems and is available from several facilities with a 24-hour turnaround at costs of $300 or less (versus a 10-day turnaround and a cost of $1800 for standard LaserVision videodiscs). Panasonic's OMDR is a nonstandard system whose 8-inch discs have dual-channel audio and can hold approximately 13 minutes of motion video or 2400 still-frames.

Note: Hitachi also makes a nonstandard record-and-playback DRAW disc system, but as yet there are no plans to market the system outside of Japan.

Eight-inch discs: Pioneer has developed an 8-inch version of the standard 12-inch LaserVision videodisc. Each side of this smaller, two-sided disc can hold either 14 minutes of motion video or 25,200 still-frames, and 28 minutes of audio (14 minutes on each channel). These 8-inch discs can be played on both 12-inch and 8-inch disc players, since playing starts in the middle of the disc and moves toward the outer edge.

Instant jump: Certain videodisc players permit the player—within specific frame parameters—to jump from one segment of the disc to another without the visual interruption, or screen "blanking," that usually occurs during searches to segments more than one or two frames away. (Depending on the player, instant-jump parameters range from 100 to 250 frames forward or backward from a particular point.) Current manufacturers of this kind of player include Hitachi, Philips, and Pioneer.

Note: A technique called *interleaving* appears to be similar to instant jump, but it actually has more to do with how the content is laid down on the videodisc than it does with player capabilities. The technique, developed and patented by Rodesch and Associates, involves "breaking up the individual alternative scenes in increments as small as video fields (two video fields make a single video frame), and then recording these increments on an alternating basis. Through track jumping during playback, a given scene can be viewed uninterrupted, or any one of many alternate scenes can be accessed within milliseconds during the vertical interval."[1]

Transmissive Videodiscs

A *transmissive videodisc* is made of a transparent material. In order to read the information encoded on the disc, the laser beam is passed *through* this material to a detector on the other side.

A recent development in transmissive videodisc technology is the Laserfilm system developed by McDonnell Douglas Electronics Company (MDEC), which uses videodiscs made of photographic film. The MDEC system, which includes a built-in compressed-audio capability, consists of a player, a recorder, a formatter, and a videodisc printer. The system is particularly applicable in situations where the information on the disc is confidential and/or needs constant updating, since the replication procedure is basically a photographic film process, and a master and many copies can be reproduced in a single hour.

Laserfilm discs are single sided and can hold either 18 minutes of motion video or 32,000 still-frames per disc; or, by using the compressed audio feature, they can hold up to 40 hours of audio.

Laserfilm discs are available only in the CAV (constant angular velocity) format (see the following description), and the disc players have instant-jump capability.

Laserfilm is currently the only transmissive system on the market. (The format was originally developed by France's Thomson/CSF, but they no longer produce their system.)

---

1. *The Videodisc Monitor*, vol. 4, issue 3 (March 1986).

Videodisc Formats

Video information can be recorded on a videodisc's surface in one of two formats: constant angular velocity (CAV) and constant linear velocity (CLV).

CAV Videodiscs

CAV is the principal interactive format. CAV videodiscs are recorded with only one frame per track. (Remember, a track is a 360-degree segment of the recording spiral.) There are 54,000 distinct tracks (or video frames) on each side of a 12-inch CAV videodisc, and each one of them could be used to display a different still-frame image. Or, all 54,000 frames could be played in a linear fashion as motion video, for a total of 30 minutes of playing time. Most interactive producers, however, use a combination of still-frames and motion video to obtain the medium's maximum potential. (For example, it is possible to have 27 minutes of motion and 5400 still-frame images on a *single* side.)

For a single side of a 12-inch disc, the capabilities of a CAV videodisc include the following:

- *Automatic stops* Some videodisc players permit chapter stops and picture stops that are accomplished by a code embedded in the vertical interval of the video frames on the disc. A *chapter stop* permits the user to access any chapter (or designated segment of the program) simply by inputting that chapter's number into the player. A *picture stop* causes the player to automatically stop on a particular frame when the player is in play mode or slow forward/slow reverse mode.
- *Dual-channel audio (30 minutes on each channel)* The two channels can be played simultaneously (for stereo) or separately (with the channels containing different information for different skill levels, or the same information in different languages).
- *Fast motion* The user can view the videodisc at rates of forward and reverse motion that are faster than the normal play speed.
- *Motion video* A single side of a 12-inch disc allows 30 minutes of playing time at normal speed.
- *Scan capability* The user can visually skim through the disc, either forward or backward.
- *Search capability* The user can instantly search to any frame on the disc by inputting the desired frame number.
- *Slow motion* The user can view the videodisc at rates of forward and reverse motion that are slower than the normal play speed.
- *Step frame* The user can step forward or backward through the videodisc, one frame at a time.
- *Still-frames* The user can "freeze" indefinitely on any one of the disc's 54,000 frames. Although the process of putting a single image on a single frame used to be expensive, facilities such as Image Premastering will now transfer single images to single frames of videotape for as little as $.50 each for 1000 images—and the cost decreases as the number of images to be transferred increases.

Note: Although all of these capabilities exist on all CAV videodiscs, not every player will accommodate them all.

CLV Videodiscs

CLV videodiscs are primarily used for linear playback applications, such as movies, musical performances, and so on. They differ from CAV discs in that they are recorded with one frame per track near the center of the disc and up to three frames per track near the outside of the disc. As a result of this structure, CLV videodiscs must play at speeds that vary from 1800 rpm near the center to 600 rpm near the outside, in order to achieve the 30 frames-per-second speed required to be compatible with NTSC (National Television Standards Committee) video—and therefore displayable on standard television sets.

The advantage of using the CLV format is that the maximum playing time of a videodisc is 100% more than that of CAV-format discs. (For example, the playing time of a 12-inch videodisc would be extended to 60 minutes.) Unfortunately, the price paid for this extended play time is the loss of many user-control capabilities present in CAV, such as freeze frames, step frame, slow motion, frame searches, and picture stops. The CLV format does, however, permit forward and reverse scanning, as well as dual-channel audio and chapter stops. Note that the CLV format is available only on reflective videodiscs.

Since CLV is relatively limited in terms of its interactive potential, our focus from this point will be on CAV videodiscs and any further reference to videodiscs can be assumed to mean CAV discs, unless otherwise specified.

Levels of Interaction

All videodiscs can be classified into one of four categories: Level 0, Level I, Level II, and Level III. (Although there is talk of a Level IV, no one seems to be able to accurately define it.) These categories were initially established in 1979 by the Nebraska Videodisc Design/Production Group as hardware definitions. Even though the ascending numbers imply an ascending order of interactivity, one should not assume that Level II or Level III is better than one of the lower levels, because each level has applications that are appropriate to its capabilities—and applications that are not.

Level 0

Level 0 videodisc players are designed for linear playback only. These players are the least expensive, but they are also the most limited, allowing virtually no user interaction. Player capabilities differ (depending on the model) but *sometimes* include forward and reverse motion at variable speeds, audio channel selection, and frame number identification.

Videodiscs produced for Level 0 players are usually straight, linear-play applications, such as movies, operas, rock videos, and exercise workouts.

Level I

Level I videodisc players have all of the Level 0 capabilities, plus search and automatic stops. In addition, several Level I players include interface ports that allow them to be upgraded to Level III players. Most Level I players also come with a hand controller.

Videodiscs produced for Level I players consist primarily of visual databases. Three excellent examples are: *The National Gallery of Art Disc,* which displays the entire collection of the National Gallery of Art in Washington, D.C., as well as takes the user on a visual tour of the Gallery; *The Astronomy Disc,* which takes the user on another kind of tour—a tour of the Universe—and which contains more than 13,000 still-frames and 80 video clips, including actual footage from the Voyager and Apollo missions; and *The First National Kidisc,* which consists of extremely inventive Level I game and entertainment applications that make this disc seem more interactive than some higher-level discs.

Level II

Level II videodiscs are controlled by a computer program that is handled in one of two ways: The program can be placed on the disc's second audio channel during the pressing process and automatically loaded into the player each time that the disc is played, or it can be input manually with the player's hand controller each time that the disc is played.

Level II videodisc players are used primarily for educational and industrial applications. They include all of the Level I capabilities, plus they have faster access time, program looping and branching capabilities, increased durability, a built-in interface port, and a built-in microprocessor that permits programmed control of the videodisc.

Level III

Level III videodiscs are controlled by an external computer. All videodisc players with interface ports can be upgraded to Level III players.

The external computer's added power provides the following additional capabilities:

- The videodisc can be controlled by more sophisticated programming.
- More than one videodisc player can be controlled by the same computer.
- Overlays of computer-generated text and graphics can be superimposed over videodisc images.
- A variety of user input devices can be employed.
- User input can be recorded and documented for testing and market research purposes.

AN APOLOGY

Because, at present, CD-I exists only in the minds of the true believers (and in the specifications identified in the *Green Book*—the CD-I standards document), it is not possible to provide the same kind of background information on CD-I as for IVD. Nevertheless, we hope that this book will provide you with a technical overview that will facilitate your successful participation in either interactive medium.

Steve Lambert and Jane Sallis

Trademark Acknowledgments

All terms mentioned in this book that are known to be trademarks or service marks are listed below. In addition, terms suspected of being trademarks or service marks have been appropriately capitalized. Howard W. Sams & Co. cannot attest to the accuracy of this information. Use of a term in this book should not be regarded as affecting the validity of any trademark or service mark.

MVP, QUEST, and VMI are trademarks of Allen Communication, Inc.

Apple is a registered trademark of Apple Computer, Inc.

AppleWorks, AppleWriter, MacDraw, MacPaint, and MacWrite are trademarks of Apple Computer, Inc.

Macintosh is a trademark of McIntosh Laboratory, Inc., and is used by Apple Computer, Inc. with express permission of its owner.

dBASE and dBASE II are registered trademarks of Ashton-Tate.

Commodore 64 is a registered trademark of Commodore Electronics Limited.

PLATO is a registered trademark of Control Data Corporation.

Cuisinarts is a registered trademark of Cuisinarts, Inc.

Touchcom II is a trademark of Digital Equipment Corporation.

DEC is a registered trademark of Digital Equipment Corporation.

CP/M is a registered trademark of Digital Research, Inc.

EECODER is a registered trademark of EECO Incorporated.

IEV-60X is a trademark of IEV Corporation.

AUTHORITY is a trademark of Interactive Training Systems.

IBM and IBM AT are registered trademarks of International Business Machines Corporation.

InfoWindow Display and Video Passage are trademarks of International Business Machines Corporation.

ComicWorks, MazeWars+, MusicWorks, and VideoWorks are trademarks of MacroMind®.

WordStar is a registered trademark of MicroPro International Corporation.

MS-DOS is a registered trademark of Microsoft Corporation.

Motorola is a registered trademark of Motorola, Inc.

PC-Graph-Over is a registered trademark of New Media Graphics Corporation.

NMG-FONTS, NMG-PAINT, and NMG-SLIDE are trademarks of New Media Graphics Corporation.

Laser Write, Space Archives, Space Disc, and VAI-II are trademarks of Optical Data Corporation.

LaserDisc is a trademark of Pioneer Electronic Corporation.

Knowledge Retrieval System is a trademark of Publisher's Data Service Corporation.

SelectaVision is a registered trademark of RCA Corporation.

Stride is a trademark of Stride Micro.

Sun/3 is a trademark of Sun Microsystems, Inc.

Sun Work Station is a registered trademark of Sun Microsystems, Inc.

MicroKey System is a trademark of Video Associates Labs.

VisiCalc is a registered trademark of VisiCorp.

V:Link and V:Station are trademarks of Visage, Inc.

Insight and Supercircuit II are service marks of Whitney Educational Services.

Smalltalk 80 is a registered trademark of Xerox Corporation.

Z80 is a registered trademark of Zilog, Inc.

1

An Overview of the Interactive Market

Rockley L. Miller

***Rockley L. Miller** is President of Future Systems Inc. and is editor and publisher of* The Videodisc Monitor, *a monthly report of news and analysis covering application, innovation, and technology within interactive video, compact disc, and related fields.*

Mr. Miller has tracked developments in videodisc technology since 1976 and is internationally recognized as an expert on the applications of videodisc technology. He has been widely quoted as an authority in the field by such sources as the Washington Post, USA Today, *the* Wall Street Journal, *and the* London Financial Times.

Mr. Miller holds a B.A. in industrial management from The Johns Hopkins University and an M.S. in management information and technology from The American University.

Who better, than this chronicler of interactivity, to start this book with an overview of the current state of the interactive arts?

The term *interactivity* refers to a program or course of instruction in which the content is controlled to some degree by the viewer. This is distinguished from the passive experience of watching a movie or other show typified by broadcast media.

At its most basic level, interactivity might take the form of responding to user input by slowing or speeding the motion of a video scene, or by playing it in reverse. It might include replaying a segment to catch something missed, or speeding ahead to skip over material already known. This type of interaction simply uses the control features inherent in the delivery hardware—ones that are available to the viewer at any time and with any program. I call this *hardware interactivity*.

At a more sophisticated level, interactivity can involve defined options that require the viewer to enter choices via some form of user interface. I call this *software interactivity*. The user interface may simply be the remote control for the video player, or, in more sophisticated systems (which usually involve a computer to control the video), the interface could be dedicated buttons, a light pen, a touch screen, or a mouse.

Software-controlled interactivity can range from simple (such as presenting options on a menu screen) to very complex (such as complete procedural simulations). Supporting hardware can range from a stand-alone videodisc player to a system including a computer, a compact disc player, interfaces, a printer, digital audio, a communications modem, and other components.

Every local information delivery format—whether it is videotape, audio tape, videodisc, CD (compact disc), CD-ROM (compact disc–read only memory), or the soon to be available CD-I (compact disc–interactive)—offers some degree of hardware interactivity, though not all players can be interfaced to a computer. Some industrial videodisc players and all CD-I players have a built-in microcomputer for interactive applications.

NOT A NEW IDEA

Although interactive media is viewed as a relatively new idea by many people, it actually represents a return to the earliest form of education and information delivery: one-on-one communication. The give and take of a two-way conversation provides an optimal opportunity for efficient learning, since both the teacher and the student are continually adapting the learning process to individual needs.

As our needs for mass education and training have grown over the past centuries, we have been forced to abandon the one-on-one relationship in favor of larger classrooms in which information is broadcast impersonally to the students. This allows us to reach and teach more people, but it diminishes our ability to truly respond to the specific needs of the individual. The greatest promise of interactive media lies in the opportunity to recapture through technology the one-on-one relationship that can efficiently respond to individual learning styles.

The concept of using technology for interactive instruction is as old as the computer. Computer-based training (CBT) and computer-assisted instruction

(CAI) systems and methods were developed in the early 1960s, as large mainframe computers evolved to support remote terminals.

With the continued evolution of computers into the inexpensive personal computers of today, these systems have become an accepted and widespread tool for teaching. However, computer-based systems have always been hampered by a lack of high-quality audio and high-resolution images to support the instructional effort.

Trainers and educators have long recognized the value of good pictures in instruction. In many cases, the value goes beyond that implied by the adage of a picture being worth a thousand words. Sometimes a good motion sequence may be worth a thousand pictures.

In an attempt to bring the power of good visual images to CBT, or to incorporate the interactivity of computers into established audio-visual technologies, a variety of hardware configurations have been tried over the past twenty years. Different combinations of randomly accessible slide systems, filmstrips, audiotape, and videotape have been tried in conjunction with each other and with computers, and have met with varying degrees of success. Most have been cumbersome, burdened by slow response times and limited flexibility in interactive design.

Film-based systems suffer from a number of problems in interactive applications. Random-access slide projectors can provide good visuals to augment an audio- or computer-based instructional segment, but their capacity is limited to that of a single tray or carousel—usually only eighty slides. The slides are expensive to duplicate, easily lost, and subject to physical damage. Furthermore, the projection systems are vulnerable to jamming.

Filmstrip systems that offer random access to a limited number of images are likewise subject to jamming. The filmstrip is easily torn by the sprocket mechanism, awkward to mount, and vulnerable to damage in handling. Filmstrips offer a further disadvantage over slides: Images are limited to those on a given strip, and the instructor is thus prevented from inserting different images.

Although most videotape recorders or players currently on the market allow the viewer to freeze an image, to vary the speed of playback in either forward or reverse, and to search for specific sections either by visually scanning the material or by watching a counter for a specific location, tape-based systems offer a different set of problems.

A major drawback of interactive tape lies in the physical linearity of the medium, which requires that a viewer wind through all intervening material in order to get from one location to another—a process that can take literally minutes on a full-length cassette.

A second major drawback is the physical contact between the read heads and the tape, which results in wear on both the heads and the tape under normal linear playback. This wear is more severe when special effects (such as freeze frame) are used, as they require the rotating tape head to remain in contact with the same spot on the tape for a protracted period of time.

Several companies manufacture interfaces that allow some videotape recorders to be computer controlled. However, sophisticated applications and market acceptance have been limited by the slow access speed and high wear factor of this format.

Similar drawbacks exist for audio tape, which has been used to provide an extra dimension in some CAI applications.

THE COMING OF OPTICAL MEDIA

In 1978, the introduction of the first optical videodisc player broke many of the barriers to effective design and implementation of interactive material. Videodisc players offer all the interactive control features of videotape and more. Because the information on a videodisc is read by a laser, there is no wear on the medium, even during special effects such as freeze frame. Discs offer two discrete, selectable audio channels that can be used for stereo sound or for entirely separate audio content (such as dual language narration).

However, most important for interactive use, information is accessed radially, not linearly. When branching from one point to another on a videodisc, you can literally skip over all intervening information and go directly, quickly, and accurately to your destination—in less than 3 seconds for end-to-end searching with newer videodisc players. It is this rapid access feature of the videodisc that has led to a maturing of interactive video applications.

Also, an inherent aspect of the radial access feature of videodiscs is that for interactive discs, one revolution, or track, of the disc corresponds to one video frame of information. Thus, in order to achieve a perfect still-frame, the laser need only be directed to reread the same track repeatedly.

The 54,000 tracks available on one side of a videodisc can store half an hour of motion video or 54,000 discrete images—quite an increase over the eighty-slide capacity of the standard carousel. And since each track has its own number and can be accessed within seconds, the sequence of images is easily modified in response to user input.

All industrial videodisc players and most consumer players are easily controlled by computers. The interfaces to do this have been on the market for six years, and the hookup and control procedures are quite straightforward. Under computer control, the player can be instructed to perform any of the functions available through the remote control. Further, with many new interfaces, the signal from the videodisc can be mixed with information from the computer and combined for display on the screen. This permits the computer to provide any volatile information (such as price in a point-of-purchase application), while the disc provides a nonchanging picture of the product.

Similarly, compact disc players can be computer controlled to provide almost instantaneous random access to high-quality audio segments. CD-ROM offers the added dimension of over 550 megabytes of rapidly accessible data.

Finally, CD-I offers high data capacity combined with television-quality image storage for play on a standardized player that includes sophisticated graphics processing capabilities. Specifications for this new format have only recently been established by Philips and Sony, and CD-I players are not expected to become available until mid- to late 1987. However, the entire CD-I concept is targeted primarily for an initial thrust into the consumer market. It is hoped that the format will be able to piggyback on the enormous consumer success of compact disc audio systems and thus open a market for interactive programming in the home. Maintaining formal compatibility, CD-I players will be able to play all audio compact discs, as well as CD-ROM discs.

It is apparent that the videodisc encompasses the capabilities of film and videotape while adding features that make it ideal for interactive applications. Similarly, CD-I encompasses all the capabilities of audiotape, with added features that place it at the cutting edge in the interactive designer's toolbox.

INTERACTIVE MARKETS

Several markets for interactive programs have emerged, and others continue to spring up, each bringing its own opportunities and special design considerations. The following discussion necessarily focuses on videodisc markets, but it is assumed that similar markets will exist for CD-I applications once they are developed.

Consumer Applications

The introduction of consumer videodisc players in December 1978 presented producers with a unique challenge to develop programming that took advantage of the videodisc's inherent interactive capabilities. One of the first companies formed to answer the call was Optical Programming Associates (OPA). OPA's first disc (*The First National Kidisc,* released in early 1981) remains a classic example of creative consumer-level interactive programming.

The *Kidisc* can best be described as an interactive video toy box for children. It demonstrated how a disc can be viewed as an information and entertainment resource, rather than as a single program. Its innovative use of all features of videodisc players, including stop action, slow and fast motion, random access, and dual audio tracks, offered hours of entertainment spread over twenty-six different segments. Subjects included making paper airplanes, learning sign language and pig Latin, performing magic tricks, and learning dances, jokes and riddles, and a variety of skill games. The creative programming techniques developed for this game-disc have seldom been equalled in a consumer product.

Another milestone came the following year as Vidmax introduced the first MysteryDisc (*Murder Anyone*). This disc used the dual audio tracks in conjunction with its video segments to create an interactive murder mystery with sixteen different plot possibilities. Patterned after the game of *Clue,* it was designed for group viewing, with teams striving to identify the who, how, and why of the crime.

Although these initial interactive titles focused on home entertainment, later interactive consumer titles have bridged the gap between entertainment and home education. The most prolific producer of such titles to date is Optical Data Corporation (formerly Video Vision), with its Space Archives and Space Disc series. Over a dozen titles provide footage from Voyager and Apollo space missions and space shuttle flights, as well as images from Landsat photographs and material on astronomy, geoscience, and a variety of other space and physical-science related subjects. In all cases, the disc serves as an archive of visual material, storing appropriately indexed collections of still images and motion sequences.

Art is another subject ripe for interactive consumer discs. Notable projects to date are a disc on Vincent van Gogh, produced by North American Philips, and the National Gallery of Art disc, produced by Videodisc Publishing. Both discs provide linear programs on one side and still-image collections of art on the other. A more recent disc, *Pearlstein Draws the Artist's Model,* presents an actual art lesson for the viewer.

Other interactive consumer titles range from *How to Watch Pro Football* to *Developing Your Financial Strategy.* Videodiscs are available to teach you how to cook, belly dance, take photographs, bowl, relax, or defend yourself—to name just a few of the many topics available.

Classic movies such as *King Kong* and *Citizen Kane* are available with second soundtracks providing running narrations on how the film was shot, and are supplemented with archives of original promotional posters, shooting scripts, and budgets.

The National Air and Space Museum offers a completely indexed disc containing 100,000 photographs of airplanes. Finally, for $89, you can buy Grolier's *KnowledgeDisc,* a complete 9-million-word electronic encyclopedia.

Consumer Design Considerations

The most obvious constraint in designing videodiscs for the consumer market is the lack of computer control. However, while interactivity is limited to the basic features of all videodisc players, it need not be unsophisticated. The key factor in consumer design considerations is to keep the interactivity simple and straightforward and to organize the information on the disc in a logical and accessible manner.

Arcade/Game Applications

The market for arcade videodiscs was a short-lived phenomenon launched in the summer of 1983 with the introduction of the highly successful *Dragon's Lair* game by Cinematronics. The market achieved orbit in the fall of 1983, as sixteen new videodisc games were introduced at the Amusement and Machine Operators Association convention and some 30,000 orders from arcade companies were placed. Manufacturers of videodisc players were put in a back-order situation for the first time since videodiscs had been introduced.

The game market then nose-dived in the spring of 1984. There were several reasons for this sudden plunge: the lack of creativity displayed in the games, the unreliability of hardware in harsh environments, and the high cost of the hardware and discs as compared with other game machines.

A few small companies continue to serve this market, however, mostly in specialty niches such as gambling casinos and bars. The most successful casino game to date has been *Quarterhorse,* which offers electronic pari-mutuel betting on a horse race that is randomly selected from the videodisc. This game broke the coin-op record in Las Vegas, with $1 million wagered during its first ninety days of operation.

Game Design Considerations

In designing arcade games, pace is all-important. Whether the game uses videodisc or computer graphics, it must be fast paced, entertaining, easy to learn, and hard to master. The fundamental rule is that great video will not make up for poor game design.

Point-of-Purchase and Public-Access Applications

Interactive systems are being used in retail and public locations to provide consistent product sales pitches, community information, banking information, electronic catalogs, and (in some instances) actual transactional capability that allows the user to purchase items for later delivery.

The first major purchase of videodisc players (and arguably the beginning of the videodisc industry) was made by General Motors in 1979, when it bought 11,000 players for use in dealer showrooms. The network, which is still in use, provides information on the GM car line to potential customers. It is also used to provide regular internal training to employees. The concept has been so successful that Ford, American Motors, Toyota, and Nissan have all followed suit with their own videodisc networks.

The first purely retail experiment was conducted when Sears, Roebuck and Company placed their 1981 summer catalog on videodisc and installed interactive kiosks in stores. By any objective standard, the experiment was a dismal failure except for the valuable information gathered. The most obvious reason for its failure was the use of a standard videodisc keypad as the means of interaction. Users were required to enter as many as five digits and then press the search key before they could get anywhere on the disc—a process that was cumbersome at best and certainly offered no advantage over the paper catalog.

Shortly thereafter, Cuisinarts™ introduced a videodisc kiosk to support sales of its kitchen products. This unit offered short video segments showing how to use different food processors and provided information on warranties and features. Viewer interaction was limited to one of several single-button selections. This concept has proven very successful, with Cuisinarts boasting increases in sales of as much as 60% to 100% at stores using the system. Over 200 units are currently installed.

As of this writing, a number of interactive retailing and public-access projects are underway. For example, several cities have created systems to provide information about shopping and business opportunities. Some supermarkets are installing kiosks that combine check approval with a quick product pitch and a coupon offer. Many national parks offer guests a brief interactive-video preview of park attractions.

In the financial community, banks are discovering the efficiencies of using information kiosks to answer a large volume of basic consumer questions on subjects such as loan types and rates, IRAs, loan qualifications, and account types. First National Bank of Chicago uses a system solely to promote IRAs, and it reports that 20% of the customers using the system subsequently opened accounts. Sears is again testing the interactive concept, this time providing information on its various financial, insurance, and real estate services. The company reports that it is happy with the initial results and is expanding the use of the network.

Levi Strauss has been testing interactive kiosks in the jeans departments of retail stores and is finding sales increases of 15% to 35%. This increase is sufficient to provide a six-month payback on the retailer's $5000 hardware investment. The company has committed to rolling out this system on a national basis.

Florsheim Shoes has been using interactive video to provide information about styles and sizes of shoes that might not be carried by the local store. Sales increases of 15% have been reported, with a substantial increase in multiple-pair sales and sales of odd sizes. Florsheim is also expanding its network.

The Sheraton and Hilton hotels have been installing interactive systems to facilitate checking in and out. Kiosks located near the front desk tie into the hotel computer and can process a checkout in less than 15 seconds. Both chains have been surprised by the number of guests using the service during peak hours. They plan to expand both the number of installations and the number of information services offered.

Point-of-Purchase and Public-Access Design Considerations

In designing an application for point-of-purchase or public-access use, it is important to keep in mind that you are dealing with a noncaptive audience of predominantly first-time users. Therefore, systems must be easy to use and must provide concise and valuable information. Users who find the systems awkward, uncomfortable, or difficult, or who find themselves locked into a long, boring video segment, will quickly walk away. Furthermore, since no one *has* to use the system, it must include some method of attracting users—either a self-running audio or video segment, or some innovative design aspect of the kiosk.

Training Applications

The one-on-one advantage of interactive video systems becomes most evident when they are used for training. Students can move through material at their own pace, reviewing difficult material without worrying about holding up the rest of the class. Thus, one benefit of interactive video training has been a shift in goals from achieving a passing knowledge of the subject to achieving mastery.

Training applications have ranged from the simple to the extremely complex. The simplest applications are straight transfers of linear material to videodisc. The advantages here are the enhanced interactivity offered by the hardware features and the durability of the videodisc as compared with tape. McGraw-Hill Training Systems has taken such an approach by transferring three of its most popular short sales-training films to one disc. Unlike the sales-training films, the disc also includes a trainer's guide, which suggests several stopping points for discussion or emphasis.

The most complex applications involve simulators, often combining several videodisc players, all under computer control, to create an artificial environment for exploratory learning. One such technique, called *surrogate travel,* was pioneered by the Massachusetts Institute of Technology, under a grant from the Defense Advanced Research Projects Agency (DARPA).

The earliest demonstration of this technique took place in 1978. Prototype players were used, since production models were not yet available. MIT filmed every road and intersection in Aspen, Colorado, and transferred the film to disc. The same disc was loaded on two players, under computer control, and a viewer could sit in front of a television screen and use a joystick to control the speed and path of travel through the town. Later enhancements allowed the viewer to switch seasons at the push of a button or to stop and enter various stores to examine their contents. The use of two players allowed for seemingly instantaneous branching, as the computer would constantly pre-cue the idle player to the next available option point and then merely switch the video source to respond to new directions.

This simulation technique has become a standard tool in videodisc applications. It is currently being used to train technicians on areas of nuclear power plants that should be accessed only during emergencies. Other applications include familiarizing security personnel with embassy floor plans and taking students to places that are otherwise inaccessible—such as the inside of a running engine.

Another major breakthrough in interactive training design came from the American Heart Association in the form of a stand-alone videodisc-based trainer for teaching cardiopulmonary resuscitation (CPR). Designed by David Hon, this system combined videodisc instruction with a computer wired to a training mannequin. Sensors in the mannequin provided feedback to the computer, which converted it to an understandable format and displayed it to the student. The experience of being "coached" by the "patient" proved to be an extremely effective instructional technique, significantly reducing the time required to acquire the necessary skill level for certification. This system is now marketed nationwide by Actronics. Several other courses have been developed using such mannequins for training in medical diagnostic procedures.

Industrial training represents the largest single use of interactive video to date. Of the approximately 100,000 systems in nonconsumer use in the United States as of mid-1986, almost half are used for training. Of these, almost half are being used by the automotive industry, where training is offered on sales, maintenance, and manufacturing.

For sales training, the videodisc represents an effective means of distributing information on new models and sales techniques to widely dispersed dealerships. The still-frame capability of videodisc permits the inclusion of numerous information- or fact-frames for easy reference.

For training mechanics, the benefits are "manifold." The constantly increasing complexity of automotive electronics requires additional training every year simply to maintain technical competence. The cost of centralized training, involving not only travel, expenses, and salaries, but also the time that the employee is away from the dealership, is also increasing. By distributing the training to the dealerships via videodisc, its cost is dramatically reduced and training is made available on demand, to fit the employee's schedule. It is also available as a continuing resource for reference and refreshers.

A variety of different types of training are now being offered on videodisc, and many companies are developing generic courses for sale on the open market. The first subject area to be broadly covered in this manner has been computer literacy. Courses are available ranging from the basics of computer concepts, to specific microcomputer software packages, to operating systems for mainframes. Companies such as Comsell, Interactive Training Systems, Active Learning Systems, and Industrial Training Corporation, offer a wide variety of computer-related courses on videodisc.

Another area that is emerging is basic skills training. The National Education Training Corporation has released a series of discs on basic electronics, industrial hydraulics, math and reading for electronics, industrial pneumatics, servo-robotics, and gauging procedures, to name just a few. Other types of basic skills training range from bankteller and telephone sales to word processing and welding.

Finally, a number of companies have developed sophisticated programs for teaching management skills. Wilson Learning's *Versatile Organization* trains managers to deal with different personality types within organizations. Digital Equipment's *Decision Point* puts the student in the shoes of a new marketing manager within a simulated company environment. Decisions are tracked in terms of their calculated impact on the bottom line.

Training Design Considerations

An important consideration to keep in mind while developing training applications is that the system should be capable of supporting different modes of learning. As the designer, it is often difficult to relinquish control of the program to the end user. Producers of traditional linear programs have a great deal of difficulty with this factor, since they are used to maintaining control and directing the course of a student through the material.

Another consideration is that the material should be engaging. This does not necessarily mean "a laugh a minute," but the material should be presented in a manner that strives to be enjoyable. Most training applications are not constantly subject to the "walkaway" potential of point-of-purchase applications—indeed, most trainees have a vested interest in completing the material. However, just because they have to watch the program does not mean that they will pay attention. Interactions should be meaningful, not frivolous, and designers should not be afraid to be entertaining.

Military Training Applications

The U.S. military has been interested in interactive videodiscs since their inception, and it began funding research projects before the technology was even on the market. The areas explored have ranged from technical and equipment maintenance training (that often offer simulations of equipment) to personnel skills such as officer training, in which videodiscs are used to simulate personal interactions.

One of the earliest actual military skill training discs was the *Army Tank Gunnery Trainer* developed by Perceptronics and demonstrated to the Senate Armed Services Committee in December 1980. Along the lines of a video game, the unit simulated the actual tank firing station and used videodisc images to provide a number of encounter scenarios. The student had to identify the target, select ammunition, aim, and fire in a real-time simulation. Shell strikes were graphically shown in the field of view. If an enemy was not hit in time, it began returning fire. The self-scoring unit provided students with both realistic experience and a competitive game-like motivation to learn.

Although this type of training has indeed been effective, the biggest benefit to the Army has come in the form of cost savings on ammunition. A single round of tank ammunition costs over $100. Therefore, the cost of live training is very high. Each simulator pays for itself during almost every training session. Recognizing this savings, the Army has recently ordered over 900 such systems, and many foreign governments are purchasing the devices as well.

Another videodisc-based simulator was developed and evaluated by WICAT in 1982 for operator and maintenance training on the Hawk missile system. Tests indicated that students using the system were able to achieve mastery of the material in about half the time required when using more traditional teaching techniques.

Given results like these, the military is one of the biggest potential users of videodisc systems for training. In fact, the U.S. Army is currently engaged in the procurement process for a new system called the Electronic Information Delivery System (EIDS). This system combines a videodisc player, a microcomputer, a monitor, and a variety of interface devices, such as a light pen, a mouse, a touch screen, and a keyboard. Further capabilities include the ability to store digital data on the videodisc and to provide sound over still images.

The Department of Defense has already declared the EIDS system as the standard for all computer-based and video-based training for all branches of the military. It is estimated that the standardization of this system could lead to over 40,000 EIDS units being installed over the next five years. Software development for the training network will likely exceed $1 billion over its life cycle. Courseware representing over 6000 student hours of instructional material is already under development for the 1986 and 1987 fiscal years.

Why is the military so excited about this technology that it is willing to make such a large commitment? The answer goes beyond cost-effective results; it borders on necessity. The military is faced with a major training problem that continues to worsen. The equipment being used in the military is becoming more and more sophisticated, and traditional training methods cannot keep up. The equipment is also getting more and more expensive, so allocating units for training purposes is often cost prohibitive. This case was well made by Col. Roy Bernd (Chief, Army Communicative Technology Office) in a December 1985 presentation about the EIDS procurement:

> Why, then, is the Department of Defense so interested in this technology? At the signal school, they are developing a repair course for the new, $4 million TTC-39 switchboard that all services will use.
>
> Presently, the school has eight real systems and can provide trainees with 11% hands-on training. However, the students need 60% hands-on training. Over 350 students must be trained each year to repair these units.
>
> In order to meet that load, the school must go to a three-shift operation, but, in order to maintain the units, they have to get down to two shifts. To get to two shifts and increase hands-on to 60%, they need 30 more systems at $4 million each, for a total of $120 million—not including the cost of a new building to house them. That's a problem.
>
> However, the school began in-house development of courseware for use on the interim EIDS system and discovered that the same goals will be accomplished for only $1.7 million, with a training effectiveness equal to hands-on experience.
>
> The overall savings of $118.3 million (on only one course at one school) is more than the likely acquisition cost for the entire 20,000 EIDS systems [the then estimated number]. This type of savings gets you funding approval when you go before Congress.[1]

---

1. "A User's Perspective," *The Videodisc Monitor*, January 1985.

Military Design Considerations

The military is undoubtedly the largest mass producer and deliverer of training. For this reason, military program design will tend toward standardization, just as its systems must, in order to support large-scale implementation. Once certain formula approaches are developed that seem to work for certain types of training, the military will tend to stick with them rather than invest heavily in exploring new strategies.

However, while the military will design a number of courses internally, a large volume of courseware development contracts will go to private industry. Many of these contracts will be issued under small business set-asides. This should prove a boon to many small enterprises seeking to establish a foothold in interactive design services.

Education Applications

The education market is perhaps the one that could most benefit from interactive technologies. Unfortunately, it will be among the slowest to develop for a variety of reasons.

The biggest barrier is the lack of a central purchasing authority. It is very difficult to market programs that require fairly high capital investments to every separate school board around the country. However, major efforts are under way as different regions have begun to form consortia to develop interactive curriculums. In addition, the National School Boards Association has taken an active role in spearheading interactive video development efforts and has formed the Institute for the Transfer of Technology to Education expressly for that purpose.

One of the first areas of development, and the subject of several projects, is basic science. The ability of interactive systems to not only present the information but simulate science laboratory experiments is very attractive to schools that are suffering from a diminishing supply of qualified teachers in this area.

Some companies are publishing generic archives of images and video footage that relate to a given topic in education. These discs can act as an instructional resource at virtually any grade level. A prime example is the *BioSci* disc from VideoDiscovery, which contains a collection of over 5000 images covering numerous aspects of biological science. An example of how this program could be used at different levels would be to ask grade-school children to identify a specific animal by name while asking graduate students to identify the Latin genus. Programs and computer interfaces to support the development of such lessons are available.

Optical Data Corporation has published an entire series of discs using footage from NASA space missions to cover general topics related to space exploration. The company offers an inexpensive interface for Apple computers and markets programs that use its material.

Ironically, the greatest use to date of videodisc in schools is not found in classrooms but in guidance offices. Info Disc Corporation and the Learning Resource Network have installed over 1500 players to provide students with information on different colleges and career opportunities. Perhaps the major reason for such successful penetration of this market has been that the players and discs have been given to the schools without charge, and the projects are funded through advertising revenues collected from the participating colleges.

Education Design Considerations

One of the biggest challenges in designing programs for education is to integrate the program into the overall curriculum. It is one thing to provide an enormous visual resource to a teacher, and quite another to inform and support that teacher in effectively using the material. Another challenge is to design material that can be used interactively in a classroom setting. It will probably be some time before many schools are equipped to provide extensive one-on-one opportunities, and most programs are likely to be used either in classroom settings or by small groups.

Custom Applications

Custom applications (a final catchall market category) are typically unique, one-time, closed applications, such as museum exhibits or trade show displays. One of the largest is at Walt Disney World's EPCOT Center, where over 275 videodisc players provide the video for a center-wide information system, as well as video sequences for many of the attractions.

One of the greatest arguments for the use of videodiscs in exhibits is their durability. Tapes and film wear out rapidly under constant use and need to be replaced every few weeks, whereas discs just keep on playing. The other major argument is the rapid access. As in point-of-purchase applications, the exhibit viewer will quickly walk away if forced to wait too long for a system to respond.

Trade show exhibitors have become especially attracted to the use of interactive systems. Such systems can be very engaging for the viewer and can also be used to collect valuable demographic information.

Custom Design Considerations

Most custom designs are similar to point-of-purchase in that they must capture their own audience. For this reason, segments must be short and easily accessible.

In trade show environments, information must be offered that is relevant and has perceived value for the viewer—not just a sales pitch. Further, if one of the purposes of the videodisc is to collect information, something must be offered in return. This may be in the form of a contest prize or simply a quick calculation that tells viewers where they fit in with the rest of the crowd.

THE FUTURE

Videodisc and CD-I represent quantum leaps in technology. They do not simply improve on previous methods; rather, they open new doors and present opportunities never before available. We are only beginning to skim the surface of the potential represented by the marriage of optical disc formats with computers. Never before have we had at our disposal the vast amount of data or the large number of images now available through this combination.

But the advent of this marriage has not been entirely smooth. We have had to develop new tools to allow video and computer displays to be merged. We have had to develop new systems to support the complex development processes involved in interactive programming and design. Above all else, we have had to learn much about how people relate to, and interact with, these systems, so that we can make the man-machine relationship as fulfilling and natural as possible.

The overall growth of the interactive-videodisc market has been slower and more frustrating than any of the pioneers would have predicted. Those who focus solely on its potential can be easily overwhelmed by the myriad opportunities it represents, and they cannot understand why everyone has not invested in such marvelous technology. Those who are more deeply aware of the components of an appropriate and successful application tend to be more conservative in their outlook. They recognize the vast potential but also realize that we are still at the early stages of the curve in learning how to design programs to capitalize on that potential. This, of course, is the key to the eventual success of interactive systems: a recognition that it is not the theory—but what you do with it—that counts.

With this in mind, it is heartening that the industry is beginning to grow at a steadily increasing rate. Sales of videodisc players for 1986 are expected to triple those of 1985. The sales for CD-I cannot be predicted at this stage, but it is worth noting that the compact disc audio player was the most successful consumer electronics product introduction in history. Close to 1 million players were sold in the United States in 1985, and that number is likely to triple in 1986.

Thus, the future for interactive media is extremely promising. Both the benefits and the needs are too great to be ignored, and, like any idea whose time has come, the progress of interactive media is now inevitable.

2

Parameters for the Design of Interactive Programs

David Hon

David Hon *is President of IXION, Inc., a Seattle-based company that designs and develops interactive applications. IXION's most recent endeavors include a software package for scriptwriters* (Split/Scripter) *and a welding simulator in which the student actually learns to weld on the screen by using a modified torch.*

Before founding IXION, Mr. Hon designed and developed the well-known interactive CPR system. He holds a patent on the IXION 100, an interactive systems controller for videodisc, videotape, CD-ROM, CD-I, and videotext.

His interactive designs are distinguished by a concern for, and sensitivity towards, the user, and you will see that concern articulated in the following chapter.

Interactive programs, whether for videodisc or CD-I, distinguish themselves from other video programs and visual or text databases by having elements that are designed not only to facilitate the user's selections but also to "engage" the user. This goes beyond mere transparency or other ease of use. Designers of most computer-based programs have learned the value of removing obstacles to the user's direct input and rapid understanding. *Engaging the user* means, quite simply, that we begin with the assumption that the user must *want* to use the program—both in the overall sense, and at each point of input.

INTERACTIVE COMMUNICATION

The best interactive programs begin with an attitude by designers and programmers that the user must be satisfied at each step and must also feel a true sense of conversational "relationship" with the program. This hidden factor in man-machine dialogues usually goes undiscussed, even though it is often the most obvious reason for the failure of programs that are otherwise operationally sound. Not only the program, but all other facets of the man-machine environment, must be selected and fine-tuned to form a communication loop in which the user participates willingly and enthusiastically.

Interactive programs, more than any other type of computer programming, must try to simulate the most human of situations: the person-to-person dialogue. This is why they demand an extremely high degree of planning and why their overall design must begin and end with the user as a participant in the system. It is also why response time in the interaction of system components is critical, though in many ways different from the sheer speed/volume requirements of data processing or telecommunications data transfer. If this simulation of human interaction is to be achieved, peripherals must be easy for people to use in a natural manner. Memory size and storage parameters must be carefully considered, not for their effect on the machine or for ease of programming, but for an overall end effect on the user—and this can result in surprising trade-offs. Interrupts, which allow the hardware to be diverted to a new task upon occurrence of some event, such as the click of a mouse button, are essential, and the necessity for certain flexibilities in system construction is a governing factor in hardware selection. And finally, size and cost must be considered as elements that may affect the usability and availability of the interactive environment. This chapter will discuss in detail each of these parameters.

Before production of an interactive program begins, a model must first be drawn in the mind of the interactive program designer: the model of the perfect interactive communication. It can be fashioned from a combination of the best parent, the best teacher, the best manager, the best entertainer, and the best friend the program designer has ever had or imagined.

This model exists in some form deep in the psyche of each human being because the ability to communicate interactively, on a range of levels from biological to metalanguage abstractions, has perhaps been far more significant than the development of opposing digits, or even a reasoning brain, as the ultimate distinction between human beings and animals. Human beings have progressed only because once they have experienced a problem and shared this experience with other humans, they eventually create a new tool to solve the problem. Thus, progress comes about only through human beings' interaction with problems and communication with each other.

The responsibilities of the interactive system and its programming are to take the best of human experience and transfer it to a man-machine experience in a manner that carries the best of humanity forward through the machine—quite often despite the machine. Concepts such as "lively," "intriguing," "fascinating," "nurturing," and so forth may not seem to belong in the vocabulary of a computer-based system. However, unless these words, which describe very human relationships, are among the governing descriptors for an interactive system from the start—not as a marketing afterthought—then that system should not be called "interactive" at all.

Beyond what an interactive program actually does, it also communicates a certain "personality." Since that personality is the reflection of the human consciousness behind the programming, program personalities vary immensely depending on the people involved. A tremendous mistake in programming interactivity would be to attempt to create a program that had no personality at all. From the viewpoint of interactivity, it is better for programs to vary between programmers, just as human relationships vary between people, than for them to seem totally sterile, with no personality whatsoever.

The designer of interactive programs must, then, be a very special kind of programmer. He or she must realize that all aspects of the program exist not only for the user's ease and convenience but also to achieve a reasonable (to the user) facsimile of human interaction—rather than asking the user to become a tool of machine interaction. This perception alone may separate the programmer of interactive media from other computer programmers. Certainly, it will separate truly interactive programs from those that the user uses only because he or she has to. And computer people are discovering that such user-unfriendly programs are easily displaced by a new generation of programs that create pleasant experiences for the user. It is an epiphany that may open the marketplace from 20 million to 200 million users within a few years.

On a more objective plane, there are two distinct kinds of interactive programs: *one-way* interactive (*static*) programs and *two-way* interactive (*dynamic*) programs.

The term *static*, or *one-way*, interactive is used when a transaction begins with the user being offered choices and ends with the user receiving the information he or she has requested. This type of interactive mode can be extremely valuable in catalogs, manuals, encyclopedias, and numerous other collections of data, all of which may include graphics and sound as well as text. An ideal example might be a bird watchers' encyclopedia, in which the sight and sounds of the birds are accessible. However, the ability to search for a particular birdcall may lie with the next generation of voice recognition.

The one-way static interactive program is used to allow the viewer easy access to a wide variety of information. In the CD-I or interactive video medium, much of this information may be visual. With such visual information, traditional database searching techniques, which are based on key words, may not be practical. Therefore, a comprehensive hierarchy, plus other very creative search methods, must be devised to give a highly visual database the rapid and variable types of access that can now be enjoyed in the solely text-based program.

The term *dynamic,* or *two-way,* interactive is used when a transaction involves, first, a demand being made on the user; second, the user's response to that demand; and, third, a new response from the program, which quite often results in a new demand being made on the user. In this way, the program "learns" about the user, even if all it can learn is how many questions the user can answer correctly out of a set number on a test. What the computer learns in even this minimal instance can make the interactive experience intensely personal for the user; and, as artificial intelligence heightens the learning capabilities of even microcomputers, dynamic, or two-way, interactivity will become even more effective.

It has been customary for artificial intelligence practitioners to concentrate on the *operation* of the program. However, with dynamic interactivity possible, it will be at least as important for them to concentrate on the *expression* of that intelligence in the communication of findings to, or in the request for input from, the user. Only if the user clearly perceives what has been decided—or what is being asked—can he or she meaningfully respond to the computer. And no matter whether the program is interacting with the user on a static or a dynamic basis, it should be designed around the user and the context in which the user will be operating. This concept is so often disregarded, or at least discounted, when it should be the place where design begins and where validation of the system ends.

THE HUMAN COMPONENT

Human factors have been an afterthought in many computer programming activities. In interactive programs, however, it is essential that a program not only operate well but also be easy for its intended user to understand or to gain value from. Otherwise, any technical merits it may have are extraneous. As an example from everyday life, if the user cannot sit in an automobile to drive it, the finest engineered motor is virtually worthless except as an intellectual exercise. And, if the engineer begrudgingly puts in a seat, this alone does not make the automobile truly functional. The user must be able to reach all of the controls. And even if the user can reach the controls, the seat must be comfortable enough that the user can sit in it for hours without back strain. And if this car is a product that will compete with other products, there must be a full range of considerations that will make the user *want* to sit in this particular automobile and drive it.

This mundane example is an understandable description of human-factors thinking as it has evolved during this century—perhaps parallel to other kinds of engineering, or perhaps as a series of engineering afterthoughts.

However, when we are dealing with an interactive program, the human factors cannot be an afterthought. In fact, the *content* of the program, and indeed the *operation* of the program, may be several links down the chain from the establishment of interaction between the user and the machine. Although certain kinds of content may give rise to further types of interaction, the initial relationship of the user to the program should be given the earliest detailed consideration.

Thus, much of the earliest thinking must consist of asking questions about the user. Here are some simple ones:

- Why is the user using this program? (Or why are seventeen various users using it, and what priorities do we give various usage?)
- Who is the user? What mental skills and prerequisite knowledge can be expected?
- What is the user's reading level? Does the user type? What kind of "computer literacy" can be expected? What is the user's attitude toward interacting with a machine?
- What are the various contexts in which the user will be operating? Will he or she be at home, at school, at work on company time or after hours? Is the user in a shopping mall or an airport? Or in a foxhole?
- What is the time frame of usage? Will this be one long sitting or several short ones?
- What is an acceptable response time if the user requests information or if the user's performance is being evaluated?
- Will the user want to take a break during certain longer segments? If so, how will he or she return to the same place, or choose to go to another place?
- Will the user be more receptive to a regimented approach or to an "open" system? Will the flexibility of an open system be confusing, or the structure of a regimented system demotivating?

The answers to these and scores of other more detailed questions should lead to one somewhat revolutionary conclusion. In an interactive program we must design the user *into* the system. The user is a component of the system and must be made as operational as any other component—and as communicative. We must, in effect, design a user.

People being what they are—self-willed and obnoxious at times, giving and compassionate at others—we will design the most compatible system if the user we design as a system component looks like a real human being. Therefore everything we do to model human behavior and human perception will have a positive effect on the human component of the system and, by this new definition, on the success of the interactive system itself.

Such a design process is not merely answering a long checklist of questions. At a fairly early juncture in planning, and at frequent steps in the whole development process, the designer's own human intuition must come heavily into play. We are each of us, after all, a very fine "expert system" on the business of human interaction. We know about analogy and innuendo, as do all humans. We know when we are performing extra tasks that could as easily be done on a machine, and we know when the machine is taking all the possible steps for us that it can.

So, designing the user as a component is no afterthought and no easy matter, for in that preliminary and ongoing exercise we define the success of interactive programs. Following that single step, if it is well thought out, the selection of equipment is fairly easy, the method of programming has more distinct guidelines, and the accuracy of content becomes a pro forma exercise.

HARDWARE

The most common situation seems to be that the hardware for a project is chosen by one group and the programming is done by another. This is the equivalent of starting a tunnel through a mountain at the same time from opposite sides, with no means of communication during the process. Having the two sides join in the middle would—we know—be mere coincidence. And yet that is the approach taken on most interactive projects: "Here is the equipment we have" (or "Here is the equipment the marketplace has" or "Here is the maximum configuration in price, memory size, physical footprint, dangling peripherals, etc."). *"Now program something to work with that system."*

If at all possible, both the program design and the hardware selection should be done within the framework of the project, with the hardware being selected at an intermediate juncture. Doing it this way has distinct advantages.

First, the true purpose of the project—what the software can accomplish for the user—becomes the guiding factor. Cost considerations can still play a part because access for a maximum user base may depend on availability of equipment, which may in turn depend on cost, or size, or even weight. However, if these decisions are made as part of the total design environment, instead of before the environment is initiated, then the chances for profitable trade-offs are immeasurably enhanced.

Second, while you are waiting, the price/performance ratio of the hardware could very well improve. Decisions made using current pricing and current technology may date the system by the time it is implemented. The approach suggested here does not demand such accurate projections about the state of technology and markets one year hence.

Third, the design process itself has an internal "learning curve." As problems are collected and solved, certain general system solutions become more apparent. Consider, for example, the videodisc course on electronics developed for the military, which used touch screens and had students "probe" with their fingers. After the early test groups completed the course, a few of them proceeded into the field, probing circuits with their fingers. The videodisc designers promptly abandoned the touch screens and went to light pens. Luckily, they had that hardware flexibility. Without that flexibility, results might have varied between 7324 touch screens gathering dust, an excellent program going unused, or a few graduates from each class merrily testing 4000-volt circuits with their index fingers.

If it seems that too much has to be known before a project begins, it is only because planning the interactive system requires that some research be done at the earliest stage and that other research be done based on those findings. For that reason, early design that is concerned both with the user and with a clear definition of his or her consequent behavior should result in a carefully designed evaluation up front. Spending time and effort to design the measures of the program's success will (a) save time in the eloquent defenses and justifications usually spent during nonimplementation of the program; (b) give all parties a clear idea of what equipment, programming, and content are absolutely necessary; and (c) if the program validates according to plan, offer distinct quantitative data supporting the use of that program in its proposed environments.

Hardware System Interactivity

Given that the user has been designated as the first component of the interactive system, other parts of the system may need to be programmed to interact in order that certain interactive capacities and effects can be achieved. The various parts of a system should integrate well into the final effect, and peripherals should not be added if they are not integral to the total design.

A good example of this situation has occurred in interactive video, where many people have tried to justify using a computer-graphic overlay capability that can add, at this writing, an additional $1500 to $2000 to the cost of each user station. This adds $200,000 to a project designed for 100 stations—not to mention the additional programming costs. Since a laserdisc can carry quite a few text messages plus very high resolution graphics in its 54,000 frames, and since it can access both more quickly than a Winchester hard disk, there are few bona fide reasons for using computer-graphic overlays. In the absence of valid reasons, clever rationalizations have to be generated to support the purchase of overlays. Here are some of those rationalizations:

> "We may need to change some things." (This could be done by switching screens.)
>
> "We may want to reflect the user's interaction on the screen." (Good as far as it goes, except excellent feedback can be given with sound and with the instant change of the video itself. Is it worth $200,000?)
>
> "We will probably find a use for it somewhere in this program or in future programs." (This means that we'll spend the money now and think of a reason for spending it later. What usually happens is that designers and programmers put untold effort into including unnecessary parts of the program that use this capability, instead of concentrating on the objectives of the program itself.)

Peripherals

There is a modern paradox with regard to peripherals, which may not be resolved soon: The less restraint you place on the user during input, or on the system during output, the more cards and cables and components you must attach. This is not simply a matter of standards. Standards in CD-I, for instance, may soon be resolved as far as data storage, retrieval, and interface to the microcomputer are concerned. But total standards for user input and system output may be long in coming. Here are a few examples:

> The user may need to input data to the system in an unusual manner. However, the mouse, the touch screen, or similar inputs based on x-y locations on the screen will not allow voice recognition or the kind of tactile input through special mechanical peripherals that can immensely enhance special applications.

Business applications and some home applications may require dual screens instead of windows in order to have twice the data available, especially when a video image and significant data are involved.

Printers often become essential when any kind of database is being used.

The powers of videotext and other modem opportunities, as well as the untapped potential of cable TV, may significantly expand the range of the system.

Recordability, as an option in the near future, requires not only computer input but, in the case of video, camera and audio input as well.

What will be needed in the way of future systems is not at all clear at this time, but already a tradition has evolved that will make standardization difficult: The more exactly the application achieves its purpose, the higher the likelihood that a special system configuration will be needed. Thus, the dominance of one type of system over another will probably never be total, though systems are likely to evolve along one of two paths.

Big-Business Dominance

It is possible that one or two large companies will dominate the market by default, simply because it is difficult both to design enough capabilities into a system for a low enough price and to dedicate enough marketing dollars to getting the word out. Trying to dominate the home market requires incredibly high risk—too high for most small companies. The home-computer concept has not been an unqualified success, and other efforts—such as laserdisc and videotext—have been so half-hearted that they have failed because of lack of adequate software and low-cost hardware. As I've said, the concept of interactivity is elusive. Most of the companies that have tried to penetrate the home market have failed miserably, not because the hardware could not be produced at a cost-effective price, but because the relationship of the user to interactive software was not understood. Traditionally, technically based companies do not reorient themselves to the lay user, or at least they drastically underestimate the amount of reorientation that will be required. In the consumer market, this technical narcissism has been fatal to a broad outreach, and, without the very high volume that follows consumer acceptance of the software, the hardware cannot be produced in sufficiently high volume to bring the price into the consumer range. So far, only videocassettes and audio CDs have achieved any degree of consumer acceptance. Creation of the "software" for both required the highest degree of talent and was extremely expensive because of high regard for the consumer. It is unlikely that consumer software capable of driving interactive hardware—either videodisc or CD-I—will be any less complex, and it will therefore be extremely expensive because of the vast amount of information and programming that will be required to attract the lay user. Thus, it is possible that only large companies will be able to undertake development of interactive programs in the near future, with the result that big business will dominate the field.

Inductive Evolution

It seems more likely that interactive systems will develop on a trial-and-error basis, even though this path is fraught with numerous risks of a lower order. Ideally, many applications will be tried, and, out of an inductive synthesis over many years, the most flexible, most desirable system will evolve. The largest risk with big-business dominance is that one company, at one point in the development of a new way of using media, will have all the answers, or enough of them, to dominate future development by setting standards. This happened with broadcast television in the United States, which is dominated by three major networks, and with the personal computer market, which is dominated by IBM. In the latter case, the company capitalized on an established need and on an established convertible software base before mass manufacturing was begun.

The existence of a critical mass of installed systems with not only standardized data handling but also standardized peripherals will have a strong influence on the individual decisions of software programmers. However, it is unlikely that such a critical mass will be achieved any time soon, and, in the meantime, there is much experimenting to be done.

MEMORY SIZE

Available memory is one of the most rapidly evolving of the technical subsets in computer technology. When using the laserdisc without LV-ROM (laservideo–read only memory), the ability to access each of 54,000 frames in 1/30 of a second to 1 second provides permutations that will make an incredible amount of graphics programming unnecessary. Thus, with a straight laserdisc application, memory requirements may not be large at all, depending on the talent of the designer. For example, a tremendous amount of interactivity can be achieved with 24K in the smallest portable computer. Only when a large current database for *swap screens* (video-to-computer and back) or for synthesized sound (as from a hard disk drive) is used does the requirement for higher resident memory increase. LV-ROM should have even more of these advantages.

CD-I, on the other hand, must pull its video images from as much as a megabyte of digitized data, and shifting between still-frames with any speed will require very large resident memory. To include enough memory that a $199 laserdisc player can display video still-frames at its maximum rate (30 per second) would seem to require in excess of 10 megabytes of resident memory to accommodate all at once the ongoing process of searching, loading, storing, and displaying. Even in a technology evolving as rapidly as memory chips, this will be difficult to accomplish quickly or inexpensively.

To achieve real-time interactivity at a consumer-marketable price, CD-I memory must also hold digitized sound, graphics, and program data for instantaneous usage. A CD-I program designed to compete in a consumer environment, then, must be of the highest quality. The program must have an enormous number of variables, which must be handled at the highest speed. To achieve an interactive relationship with the user, a tremendous number of digitized elements must be available *in real time*.

Or almost in real time. And here is where design strategies and disc-geography planning become crucial. The user may be able to wait up to—but not more than—2 seconds and still consider his or her experience a real-time experience. Dr. Robert Fuller, in studies during the 1960s with computer dialogues, determined that 2 seconds seems to be a "waiting" threshold for humans, before their minds flit to something else. An increasing body of experience with computer-aided instruction and with interactive video seems to support these studies. In other words, if the access time is under 2 seconds, no "please wait" or other obnoxious phrase is needed. This factor may end up being the determining factor in system requirements for real-time interactivity, which will give a boost to designers who despair that the hardware is too slow.

STORAGE PARAMETERS

Even when high interactivity is the purpose of a system, different parameters are necessary for different media. What kind of data, how much data, and how quickly can that data be located are concerns that differ greatly for CD-I and in interactive laserdisc. It is here that the two media show their fundamental differences: CD-I is a constant *linear* velocity (CLV) medium that uses digital data exclusively, and videodisc is a constant *angular* velocity (CAV) medium that has superior facilities for analog video—including still-frame video—and superior speed of access (Winchester and floppy diskette systems are CAV).

Ideally, designers should have the option of deciding to use CAV videodisc if the database is largely visual and needs very high speed retrieval. Or, if the program contains a large database or music base with limited visual requirements, then CD-I might well be the medium of choice. Unfortunately, hardware systems do not always evolve around their software needs. CD-I is being increasingly thought of as a still-frame video medium because of market considerations that link it to compact discs for music and to the reservoir of data that is CD-ROM.

The laserdisc contains 54,000 frames, which may be thought of as concentric circles. In current systems, any frame is accessible in less than 2 seconds. Also, *instant jump* capabilities available in certain players allow the player to access any frame within 250 contiguous frames in either direction in 1⁄30 of a second. Analog storage of video frames is about fifty times as efficient as digital storage, far faster to access, and just as accurate. However, storing digital data in an analog signal is more difficult than storing the same data in a digital signal, and digital data stored in an analog signal is far more error-prone. Error rates of 1 bit in 10^{13} are common in a digital signal, but quite difficult to achieve in an analog signal. Of course, the video picture can bear a great deal of bit dropout before any difference in quality is noticed, but the same is not, unfortunately, true of digital program data. Thus, digital data in an analog signal must have high redundancy to achieve any acceptable data integrity.

The CD-ROM format, which is sometimes thought of as a stream spiraling outward from the center of the disc, has a very regular data compaction and thus can have very low error rates with comparatively little correction. Its appropriateness for applications involving large databases and delicate program data is therefore obvious. For the present, digital video images must be used sparingly with any CD-ROM based format, because of the very large amount of information needed for a digitized video picture. In raw form, the digital video picture requires about 1 megabyte per frame, though coding schemes will increasingly reduce this demand as CD-I becomes more standardized.

The LV-ROM format will, ideally, combine the speed and visual efficiency of videodiscs with the digital data compaction capabilities of CD-ROM. In this hybrid, 300 megabytes of CD-ROM data are carried in a low band that does not interfere with the 54,000 video frames or the two tracks of analog sound. As such, this medium could well evolve as the medium of preference when several kinds of interactive applications are required. In many cases, existing CD-ROM databases and videodisc visual databases may find a common home on LV-ROM if the CD-ROM database does not exceed 300 megabytes.

SPEED

Speed in an interactive system has several meanings:

1. Search speed through an optical database.
2. Operational speed of component interaction.
3. Perceived waiting time by the user.
4. Operative speed in RAM.
5. Responsiveness of user inputs.
6. Ease of operation by the user.

It is obvious that the second, fourth, and fifth meanings have a great deal to do with the hardware selected, and that the first, third, and sixth meanings depend more on the design of the program. However, the key factor in speed—when viewed as an independent variable—is whether the CLV or the CAV mode is used. Perhaps ironically, then, the CLV (digital) mode ends up being slower than the CAV (analog) mode. The CLV mode is continuing to evolve sector and frame recognition methods that will increase its search speed, but it may never equal the slowest CAV systems because of the CAV system's inherent edge in having frames and frame accuracy. On the other hand, when data from the CLV system finally does arrive, it can have far greater accuracy; and, if bit accuracy is the goal, the need for high error-trapping redundancy reduces the CAV speed advantage. For some data-intensive applications, then, the seemingly slower CLV systems may win the race.

The prime factor in speed for either CLV or CAV is the planning of potential itineraries through the vast available geography of the disc. Thinking of the disc as a continent of information, end-to-end (coast-to-coast) travel time is perhaps the first measure of an airplane's speed; but, with some experience, the traveler learns that the number of stops and the flight path's deviation from a straight line are significant factors in making the swiftest trip. Taking the Concord, then, may not be the fastest way to Joplin, Missouri.

The same is true when creating an interactive program. Creating the informational territory on the disc and planning search routes through that territory become a hallmark of good design. In fact, to the user, the amount of time it takes to learn to operate the system is ultimately not as important as the actual speed of operation.

This is why, even in CD-ROM databases, the way a user's mind works when searching is the key to designing the search method. Tree hierarchies may be one way to search, and key words may be another. Natural-language entry is also blossoming as a way of using artificial intelligence to assist the user. Graphic selection methods may evolve to a virtual "paint-search" technology; on the other hand, a great many users would be most comfortable, but perhaps not most efficient, if they could browse as they might in a bookstore. For its part, the computer demands some kind of order, and one of the true challenges of database search design will be to accommodate the vagaries of human order—or disorder.

It is not impossible to imagine future searches through digitized video for predominant color combinations or for patterns in music, but it may take quite a bit of learning on the system's part for it to match a human's mood. It is not out of the question, however, that the user's brain waves and heartbeat rates could be scanned upon entry to a house, so that he or she is welcomed by flat-screen hi-res video wall hangings and loudspeaker music that perfectly matches his or her mood. Imagine the surprise of a burglar, upon breaking and entering, at being greeted with the music of *The Sorcerer's Apprentice*.

From the viewpoint of mechanics, the speed of both CLV and CAV systems will undoubtedly increase as the need for the rapid random access required to achieve interactivity becomes more acute. Already the end-to-end access speed of the laserdisc has been reduced from about 6 seconds to 1 second on some players. This has been achieved partly through a new generation of solid-state laser diodes, which are much lighter, more durable, and more precise than the "gas boat" (helium-neon) mechanism of the early players. In those early transport mechanisms, speed was equivalent to torque, and the industrial players had to have far more powerful electric motors to achieve their random access speed. Lower-cost consumer players of the early 1980s had end-to-end access times of 12 to 14 seconds because of the use of less expensive, less powerful transport motors. Given the current mechanical transport times, we must admit that the Age of Transportation is not quite finished, and the Age of Information is not yet in full swing.

There are other semimechanical, semioptical factors, such as the *tilt* of the laser mirror, that can also add speed. In some current laserdisc machines, this results in a 1/30 of a second instant jump, which can occur within 250 frames in either direction. For the user, this means that the picture will change with no perceivable screen disruption. In one current application, IXION has built a welding simulator for the Academy of Aeronautics at La Guardia airport. The simulator uses a light pen "torch" with which the user can "weld" on the CRT screen. The puddle of molten metal on screen grows hotter or cooler as the torch moves closer or farther away. This light-as-heat analogy portends other kinds of reality modification through the creation of parallel realities that change more rapidly than the eye can see.

A digital video image would, of course, handle this welding simulation differently by holding the image in RAM and modifying portions of it by algorithm as the torch was moved in three dimensions. The welding simulator is an instructive example for two reasons: It shows the basic difference between data manipulation in CD-I and frame manipulation in interactive video, and it indicates that there may be areas where speed of access can be enhanced in either the CLV or CAV modes by opto-mechanical means. Obviously, other directions to be explored will be multiple lasers and two-sided read heads.

Getting to the data mechanically, then, is one evolving variable of speed, but understanding what data the user wants or needs, and knowing where it is located, will probably still be the intellectual challenge of all interactive media.

SIZE AND COST

Most programmers will tell you that programming can be comparatively easy on a VAX or on many mainframes, because there is so much margin for rework. In the very small, very low-cost systems that must evolve for either CD-I or videodisc to enter the home consumer market, the term *elegance* becomes meaningful for both hardware and software designers. The home system of choice should probably cost no more than $1200 and will probably bottom out around $500. This means that the computer that provides intelligent control of the optical disc will probably represent half the cost of the system. So let us say that each component will cost $250 and that the quality of the video monitor will be a further discretionary item for the user.

The chances are great that manufacturers of both CD-ROM drives and laserdisc players are targeting that cost figure for the home market. It is apparent that the overall size and the complexity of the connections will be guiding parameters for the home market, as well as for the business and school environments, where space has a more definable cost (although information carries an added value in the latter environments). Both CLV and CAV formats will be guided by those parameters, but their development demands will be totally different.

CLV manufacturers will already have a large installed base of upgradable compact disc players. This is seen as a major advantage to CD-I at the moment. On the other hand, the limitations on the number of video still-frames and on instantly accessible video motion may be the Achilles' heel of the CD-I effort in the home market, since consumers have come to expect motion video of at least the quality they see on their television sets.

The CAV videodisc people have already confronted the size parameter with the introduction of the 8-inch videodisc, which will carry 25,200 video frames or 14 minutes of motion—quite enough for designers who feel that videodisc is a still-frame medium and quite enough for the adept interactive video designer who knows how to intersperse the attributes of motion. What the CAV manufacturers will have to come to grips with is the difficulty of placing data in the analog signal. This may or may not be an Achilles' heel.

As far as equipment goes, it is now possible to imagine a system that can, through the economics of volume production, reach an acceptable price and size *without* the Achilles' heel of no motion video or low data accuracy. One core component medium would be an 8-inch LV-ROM player that would also be capable of playing current compact discs. Most of the actual technology for this combination has already been demonstrated. The other core component would be an on-board microprocessor that would, with a simple user x-y input such as a mouse or a drawing pad, provide the user with a *run-only* computer to access and interact with data. There is no question that such a component could be built right now if it did not require a digital video element.

This ideal system would accommodate about 100 megabytes of data, *plus* 25,200 video still-frames with computer-graphic overlay and still-frame audio capabilities. That combination of digital data and analog video would allow infinitely more visual permutations than the straight 600 megabytes on a CD-ROM system, and it would have much faster access because the different types of data would be stored in the most appropriate format.

A CAVEAT

There is, however, a final caveat even to this ideal and quite achievable LV-ROM system. And that is a warning about interactivity itself. Perhaps it is now time to bring in the anthropologists to study how modern society acts at work and at home, because there seems a fair probability that "interactivity" as a mentally active—rather than mentally passive—process may indeed be far more successful outside the home. Interactive manipulation of information in all forms is becoming a large part of the mentally active work and educational environments.

If we observe the last decade, interactivity in the home should be somewhat suspect by now. The major home-electronics booms—broadcast TV, audio systems, VCRs, and now compact discs—have all been passive media. Attempts to introduce electronic interactivity in the home, despite an investment of hundreds of millions of dollars, have finally failed. Home computers, game machines, and various videotext systems seem now to be ill-conceived experiments or fads that faded. It is true that the French are using low-cost, teletext phone directories (often for computer dating—ah, the French), so every failure deserves reevaluation in newer contexts. However, most evidence clearly points to at least this warning: The modern home may be a place for relaxation and human-to-human interactivity. It is all well and good to say that "people will be reeducated by the new media opportunities" or that "the quality of the software will be the key factor in the acceptance of interactive media" (indeed this writer has said those very things), but we would be extremely wise to consult the anthropologists now. And we would be extremely wise to put them on the payroll of someone other than the hardware or software companies.

Interactive media can work astonishingly well *if* the basic environment is conducive to its use. However, if that environment is in many ways unconducive, not even massive investment can prevent a spectacular failure that will also kill other legitimate uses.

3

Interactive Design Strategies

Steve Holder and Rod Daynes

Steve Holder *is Vice President for Applications Development of IVID Communications. He is the coauthor of three programming textbooks and has been involved in the design and programming of several interactive video training programs for Fortune 500 clients.*

Rod Daynes *is President of Pacific Interactive, an interactive consulting, design, and production firm in San Diego, California. He has been designing and producing interactive video projects since 1977. In addition, he founded and, from 1979 to 1983, directed the Nebraska Videodisc Group. Mr. Daynes is coeditor of* The Videodisc Book: A Guide and Directory *and is the author of over twenty-five articles and papers on interactive video.*

Interactive designers have the dual responsibilities of understanding the nature and variations of the dialogue (interaction) between the user and the system to the greatest extent possible, and of incorporating this knowledge into the programs they develop. Media such as the videodisc and CD-ROM provide a means of adding a new dimension to this interaction to a degree never before possible. Combined with the speed and power of computers, these media can provide the student with information and advice from the best teacher(s) available in a given field and/or enable the user to simulate events in the real world by using "real world" images and sounds. This responsibility should not be taken lightly.

What we are hoping to accomplish with the information in this chapter is to show that there are several alternative interactive feedback strategies and that, by using these (and other) alternatives appropriately, programs can be more effective, friendly, clever, and fun; and they can be decidedly more interactive.

We start with some basic guidelines. Then, beginning with the simplest of interactions, we explore variations that will gradually increase in complexity.

The examples presented here are not intended to be exhaustive, but merely to represent a variety of types. With an understanding of some of the basic elements and the variations that can be made with them, one can begin to recognize and create an almost infinite variety of types of interaction.

One more comment: While we assume that these strategies will be used primarily with *Level III* videodisc systems (computer-controlled video/audio with computer-graphic overlay and touch screen), many of these ideas can, and should, be incorporated into *Level II* (programmable industrial videodisc) systems, as well as interactive audio systems.

SOME GENERAL NOTES ON INTERACTIVE DESIGN

The following sections explain some guidelines that will be useful to remember as you read this chapter and as you design your own interactive strategies.

Timeouts

A *timeout* is a form of feedback that disappears from the screen after a specified period of time, without requiring any student input. A timeout is appropriate only for short, nonsubstantive messages, such as "Correct," "That's right," "Not here," and "Try again." A timeout should never cause anything substantive to be irrevocably erased from the screen.

Number of Possible Responses

The number of responses to a question (or the number of choices on a menu, or the number of unique inputs allowed on any given graphic) should be eight or less. There are few types of interactions that can effectively handle more than eight possible responses, and good screen design generally enforces about the same limit.

Obvious Touch Targets versus Not-So-Obvious Ones

For users of touch screens, a distinction should be made between interactions in which touch targets are obvious and those that are not so obvious, because it makes a difference in the type of interaction selected. It may be appropriate in some instances to outline video targets that are not readily apparent, thus providing the designer with a wider selection of interaction types.

Tailored Negative Feedbacks

Unique "tailored" feedbacks written for every possible incorrect response are an effective instructional strategy. They are also expensive. Use tailored feedbacks judiciously. They can expand the scope and cost of a project rapidly. On the other hand, tailored feedbacks can guarantee a higher degree of instructional effectiveness. (The reader is advised to study the concept of *adaptive testing,* a subject that, unfortunately, cannot be covered here. Adaptive testing is a means of isolating the instructional problems of a given student and tailoring feedback based on those problems.)

Correct Feedback

The student should not be put in the position of wondering whether or not a question was answered correctly (or, similarly, wondering if the menu selection actually took him or her to the place that was selected). Some kind of acknowledgment should be provided for all correct interactions. And, the feedback should not always be text. For example, touching a switch and observing that the light above it comes on might be sufficient feedback in some cases.

Associating Response with Feedback

It is valuable for some form of association to be created between an incorrect choice and the feedback, most particularly if feedbacks are tailored to specific wrong responses. The message "That's not right because . . ." is more valuable if we have identified what "that" is—by drawing a box around it or by substituting relevant text for "that."

Identifying Areas Touched

Again, for users of touch screens, it is a good idea to identify the area touched by using boxes or highlights. This eliminates confusion. If problems exist with touch panel registration, for example, they will be immediately evident, and the student will then have an opportunity to have them corrected or to compensate for the problem.

Negative versus Constructive Feedback

Why emphasize the negative? Rather, accentuate the positive. Instead of "No, that action is not appropriate at this time," why not "No, that would be appropriate only if the power switch was off"? If we think more in terms of positive and constructive feedbacks, we will improve our students' attitudes.

Control Line

It is useful to incorporate a *control line* into the program, either on the screen or with function keys (or both), which enables the student to override the action. Generally, such keys are operational in nature—for example, "Step Forward," "Scan," "Pause," "Previous Menu," "Main Menu," "Bookmark," "End,"—but they can also be more program-specific—for example, "Help," "Presentation," "Example," "Practice," "Test," "Advice," and so on. Operational functions can be incorporated fairly easily into a program, but they should not be confused with program-specific functions.

INTERACTIVE FEEDBACK STRATEGIES

The following pages describe the characteristics, advantages, and disadvantages of several interactive strategies. Each one is flowcharted. In addition, the appropriateness ("use when") of each strategy in a given situation is described. The following chart shows the symbols used in the flowcharts.

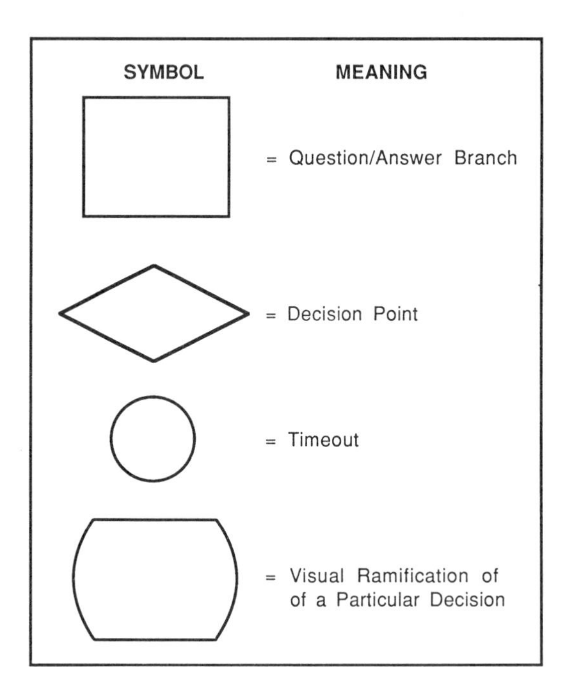

One Chance—Either/Or

Fig. 3-1 shows the preferred interaction when the student has a choice of two possible answers. If the student gets it wrong, there is no point in asking to "try again," since it is obvious that the other answer is correct. The strategy here is to tell the student *why* the answer is wrong, or tell the student *why* the other answer is correct, and move on.

Fig. 3-1. One chance—either/or.

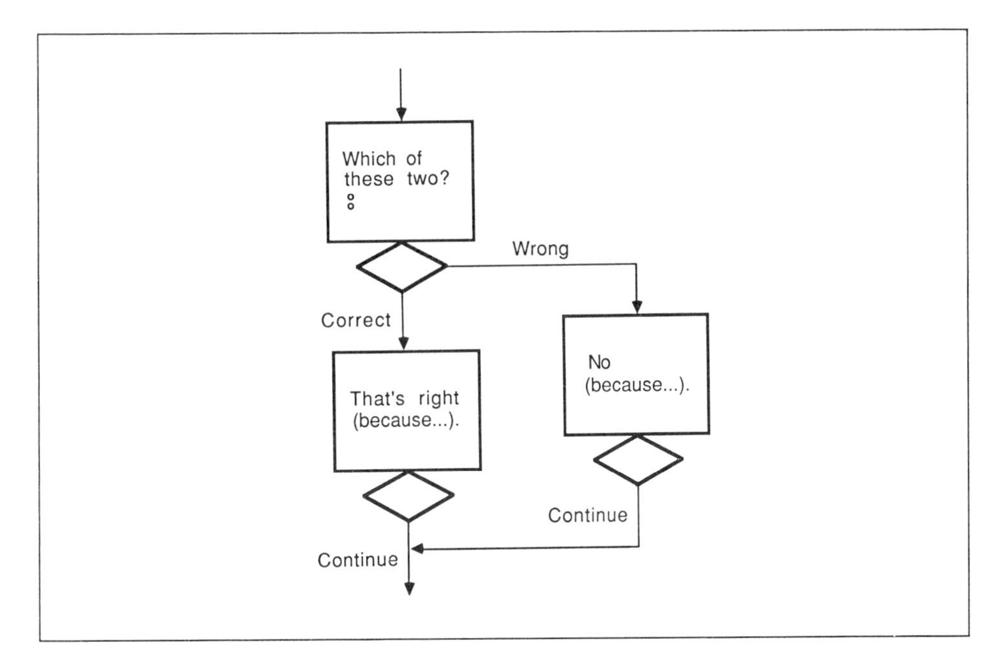

Characteristics

Classic Skinnerian use of positive/negative feedback.

The question has only two possible answers.

For each answer, a specific feedback is provided.

Advantages

Economical—requires development of only three elements:

1. The question.
2. The positive feedback.
3. The negative feedback.

Simple—branching is extremely straightforward.

Disadvantage

Boring—if too many of these interactions occur one after the other, the program can become tedious.

Use When

The two possible answers are obvious; that is, if the student gets the wrong answer, he or she can immediately know that the other choice is correct. This strategy particularly applies when the answers are text-based.

Timed-Out Correct Feedback

Fig. 3-2 uses the interaction from Fig. 3-1 to illustrate timed-out correct feedback. Correct feedback could be timed out for any interaction types that follow.

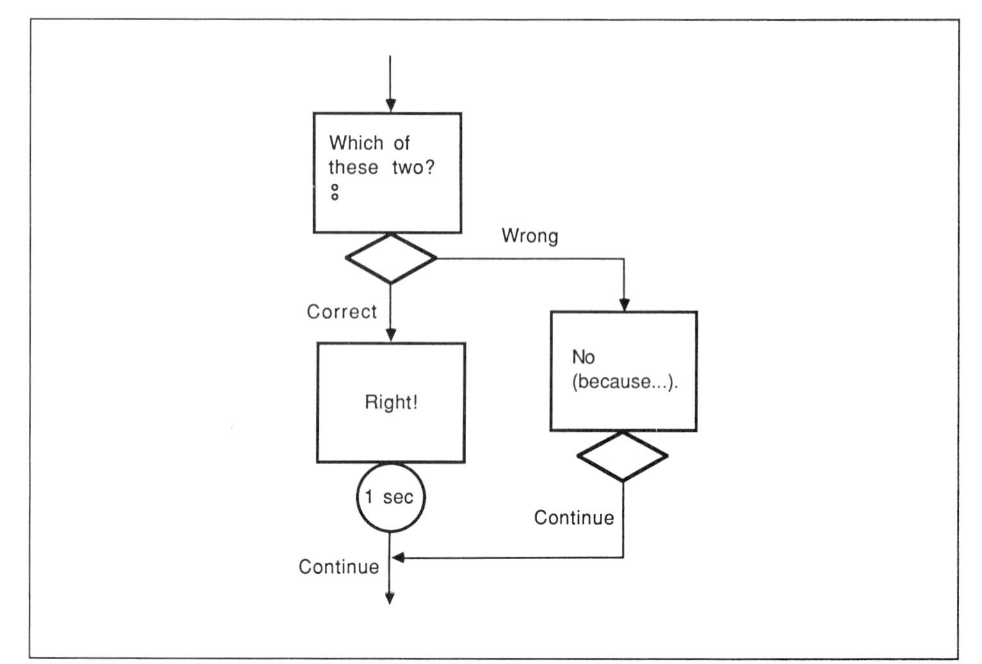

Fig. 3-2. Timed-out correct feedback.

Characteristics

Feedback disappears after a short (1- or 2-second) interval, and the program continues automatically to the next interaction.

Advantages

Eliminates an unnecessary touch when the feedback has no content and the student has no need to continue studying the question and its correct answer.

Allows the program to move at a faster pace.

Disadvantage

Removes control from the student. May cause the program to advance when the student really wanted to study that particular question or answer longer.

Use When

The feedback consists only of "Correct," "Right," "Good," or some similar short positive reinforcement. If the positive feedback has any content, timeouts should not be used.

The student does not need to examine the question and the correct answer to be satisfied as to why the answer was correct.

Do Not Use When

There is the possibility that the student might have guessed at the answer, and the feedback contained clarification.

The student might be unsure why a particular answer was correct. If this possibility exists, the questions and answers should be left on screen so that the student can examine them, raise questions with the instructor, record the question for his or her own research later, or be satisfied that it is indeed correct.

One Chance—Multiple Choices

This next strategy is structurally the same as that in Fig. 3-1, but it is conceptually different because the question has more than two possible answers. As shown in Fig. 3-3, a significant difference here is that the negative feedback has to address which answer was correct and why. Since any of several incorrect answers could have been chosen, the negative feedback cannot address why the answer selected was incorrect.

Characteristics

If the correct answer is selected, positive feedback is given.

If an incorrect answer is selected, the student is told which answer was correct.

Every answer that is not correct leads to the same feedback identifying the correct answer.

Advantages

Economical—it requires development of only three elements:

1. The question.
2. The positive feedback.

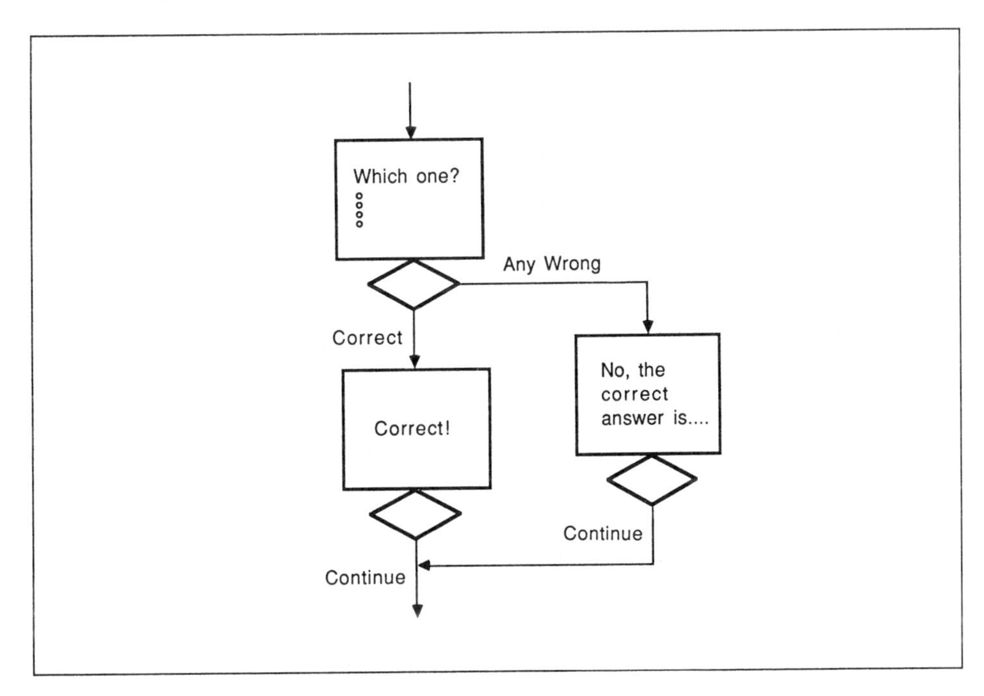

Fig. 3-3. One chance—multiple choices.

3. The negative feedback.

Simple—branching is extremely straightforward.

Uncomplicated—makes fewer demands on the student and could result in lower frustration level.

Disadvantages

Does not afford the opportunity to dispel any misconceptions that may be associated with certain of the wrong answers; that is, negative feedbacks are not tailored to specific wrong responses.

Not the most effective instructional strategy in some cases (since it caves in and gives the student the answer at the first wrong try).

Use When

Tailored feedbacks for each incorrect response are not deemed necessary.

It is important to minimize student frustration level.

Do Not Use When

It is important to identify and dispel misconceptions that may be associated with particular incorrect responses; that is, when tailored feedbacks are needed.

It is instructionally important for the student to make more than one attempt at any answer before the correct answer is given. (See Fig. 3-6.)

One Chance—Tailored Feedbacks

Fig. 3-4 shows a variation of Fig. 3-3 that provides individual feedbacks for each of the incorrect responses.

Characteristics

More than one incorrect response to the question is possible.

Each incorrect response leads to a different feedback that explains why that response was wrong and what the correct answer is.

Advantages

Makes the computer appear "intelligent" when each possibility is covered by a different response (gives the student greater faith in the instruction).

Can enable a greater degree of instructional effectiveness when specific misconceptions can be identified and dispelled.

Disadvantages

Requires creation of separate feedbacks for every wrong response (not economical).

Not the most effective instructional strategy in some cases (since it caves in and gives the student the answer at the first wrong try).

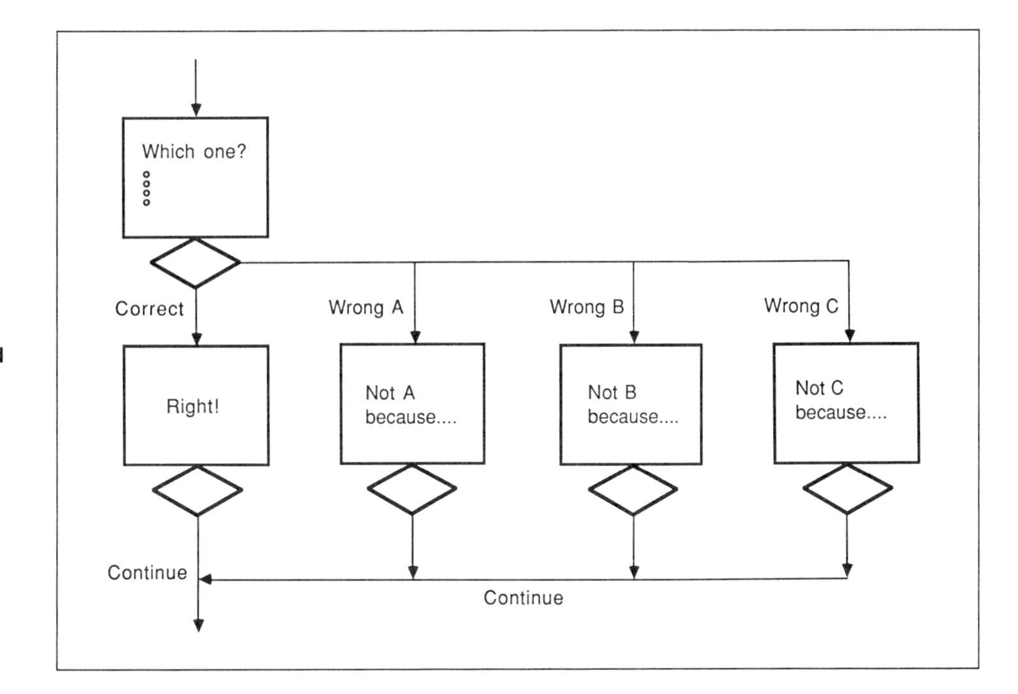

Fig. 3-4. One chance—tailored feedbacks.

Use When

Tailored feedbacks for each incorrect response are important.

It is important to minimize student frustration level.

Do Not Use When

Development cost is an important design issue and the need for tailored feedbacks is not overwhelming.

It is instructionally important for the student to make more than one attempt at any answer before the correct answer is given.

Nope, Try Again

Figs. 3-1 through 3-4 are examples of the "one chance" variety, where branching is always forward. Fig. 3-5 is the first example of a loop, which branches backward to create a second chance at the question.

Characteristics

Selecting the correct response leads to the positive feedback.

Any incorrect response causes "Try again" (or some similar message) to flash on the screen for a second or two and then disappear.

Advantages

Economical to design, since it has only three elements:

1. The question.
2. The positive feedback.

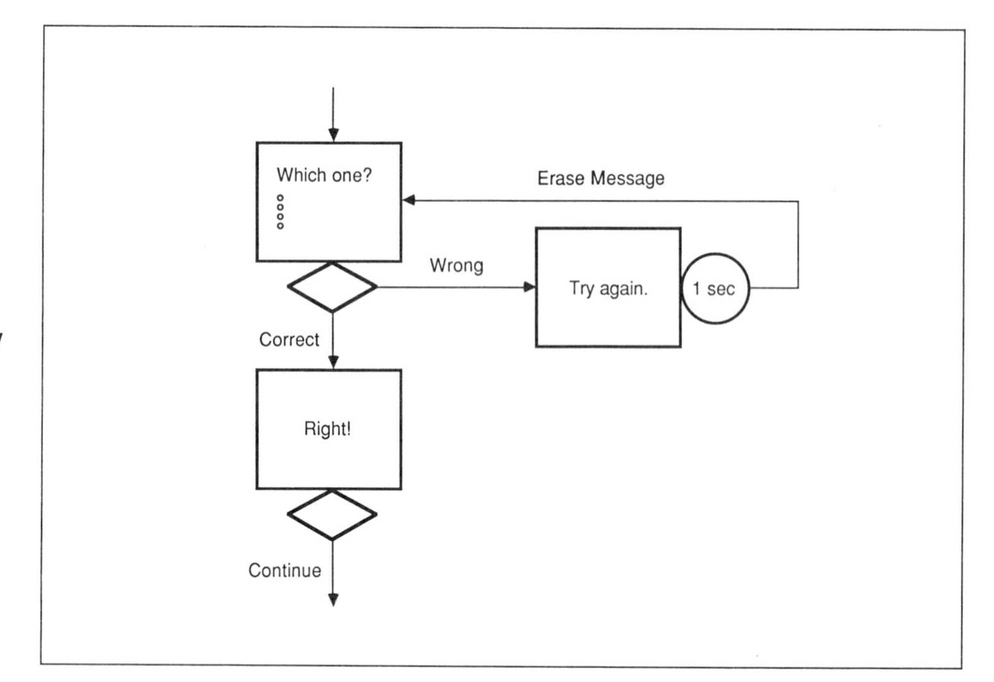

Fig. 3-5. Nope, try again.

3. The negative feedback.

And, the negative feedback does not require special wording.

Visually effective.

Can be an instructionally effective situation, since it forces the student to select the correct answer rather than telling the student what the correct answer is.

Disadvantages

Creates the possibility of an endless loop.

Provides no specific feedback for wrong answers.

Potentially very frustrating; seldom recommended for remedial students.

Use When

Tailored feedbacks are inappropriate and you have to stretch to come up with even a single negative feedback that is not inane.

There are three to five possible responses.

Do Not Use When

Only two answers are possible and each one is obvious (the "Try again" strategy is not appropriate).

Constructive feedback is appropriate for an incorrect response.

There are six or more possible responses (potentially frustrating to get nothing but "Try again" for, in some cases, five consecutive answers; see Fig. 3-6).

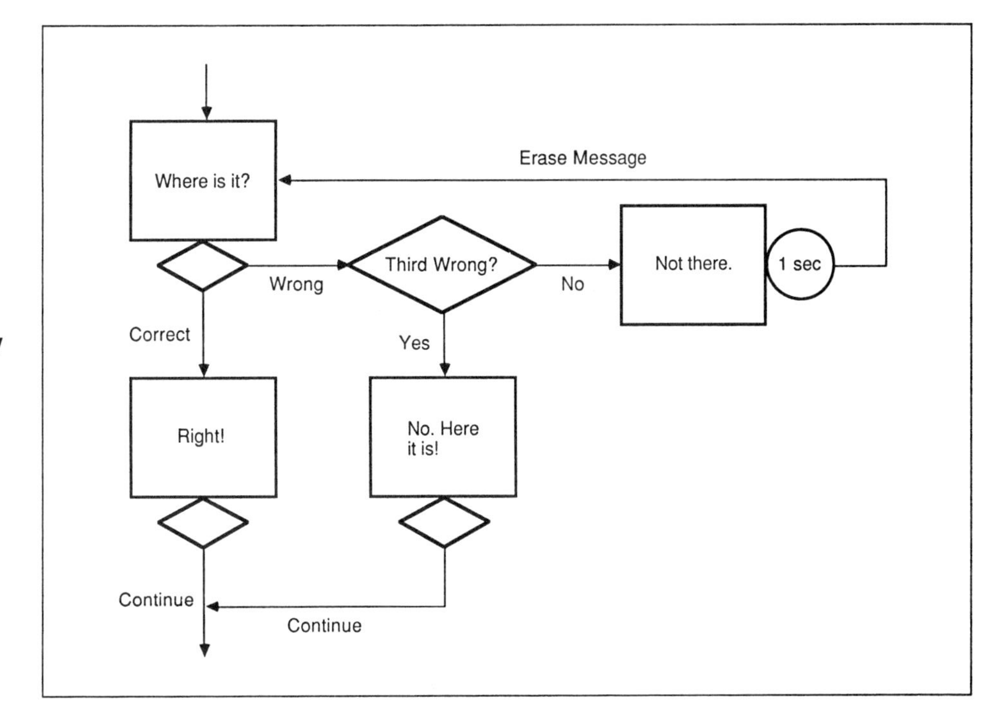

Fig. 3-6. Nope, try again (with counter).

The possible responses are not obvious (particularly when responses are a few video objects in a field of many; see Fig. 3-6).

Nope, Try Again (with Counter)

In Fig. 3-6, the interaction is a variation of Fig. 3-5 that solves the endless loop problem by placing a counter on the wrong responses. (The use of counters is expanded on in later interaction types. It is introduced here because it is an obvious fix to the main drawback of Fig. 3-5.)

This interaction can be viewed as a composite of Figs. 3-3 and 3-5, which combines the best features of each interaction type and mitigates some of their inherent weaknesses. Basically, this interaction says, "Try it yourself a few times. If you don't get it, I'll tell you."

Characteristics

Selecting the correct response leads to positive feedback.

The first few times an incorrect response is selected, "Try again" (or some similar message) flashes on the screen for a second or two and then disappears.

After a predetermined number of incorrect responses, the student is given the answer.

Advantage

Can be instructionally effective in the right situations (it encourages students to make a few attempts on their own, but does not leave them alone).

Disadvantage

Does not provide specific feedback for individual incorrect responses (no tailored negative feedbacks).

Use When

An alternative to Fig. 3-3 is required; that is, it is felt that the student should be given more than one chance at the question.

The student will require a way out of the loop after a certain number of tries.

Do Not Use When

Only two answers are possible and each one is obvious (the "Try again" strategy is not appropriate).

Touch to Try Again

Fig. 3-7 is similar to Fig. 3-5, except that the feedback does not time out. The student is required to touch the screen to receive a second chance at the question. Note: This interaction is often used where it is not appropriate.

Characteristics

Selecting the correct response leads to positive feedback.

Selecting an incorrect response gives a generic negative feedback and provides a mechanism for getting back to the question for another attempt (i.e., specific instructions to touch the screen, or a touch area labeled "Try again").

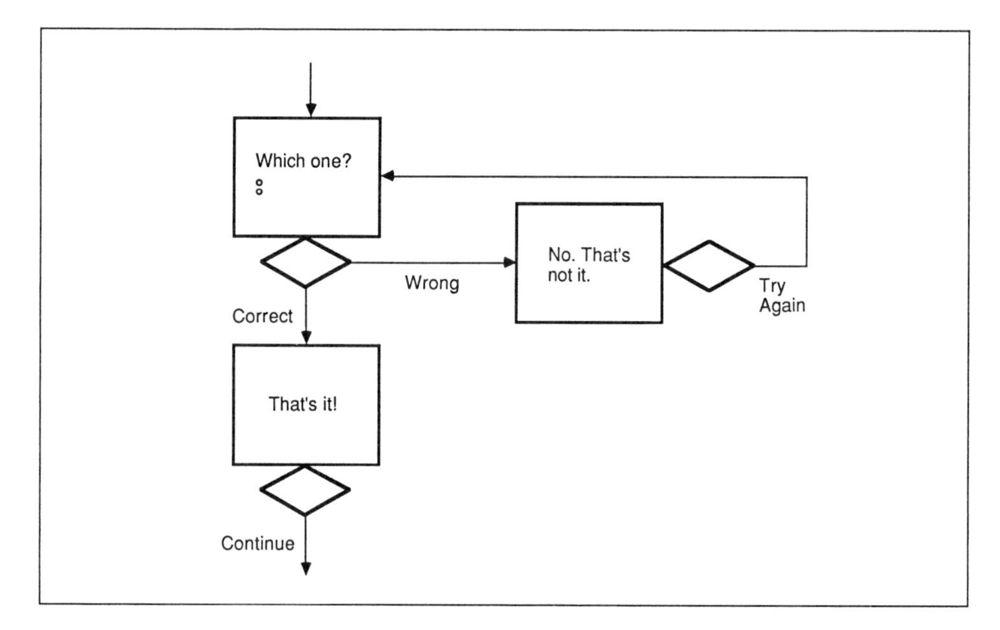

Fig. 3-7. Touch to try again.

Advantage

Provides a mechanism for a repeated attempt at the question when the negative feedback destroys the original question or answers.

Disadvantages

Does not provide for tailored feedbacks (unless there are only two choices to begin with, in which case "Try again" is not appropriate).

Easily misused (sometimes, unnecessary touches are created and if the user *doesn't* touch the screen, answers are left on the screen that are not active).

Use When

Video takes up so much of the screen that the only way to provide feedback and still see the video is to overwrite (erase) the original question (and then only if it is not possible to provide the feedback and a restated question).

The feedback for the incorrect responses consists of a different video (in which case, the question should be erased to prevent visual confusion).

The feedback for the incorrect responses consists of a dramatic video and/or graphic response that erases the question.

There are three to four possible responses and they are obvious (so the student can get to the correct answer, if only by process of elimination).

Do Not Use When

Only two answers are possible and each one is obvious (the "Try again" strategy is not appropriate).

There are more than four possible answers (could require as many as nine touches to get the correct answer).

Feedback can be written on the same screen as the original question and answers, without altering the original video or graphics.

Constructive feedback is appropriate for an incorrect response.

The possible responses are not obvious (particularly when responses are video objects in a field of many; see Fig. 3-6).

The original question must be erased to make room for feedback, unless the original question can be restated as part of the feedback.

Touch to Try Again—Tailored Feedbacks

The strategy shown in Fig. 3-8 varies from that of Fig. 3-7 in only one way: Individual negative feedbacks are provided for each wrong response. Otherwise, Fig. 3-8 has the same potential for misuse as Fig. 3-7.

Characteristics

Selecting the correct response leads to positive feedback.

Each incorrect response gives a unique negative feedback and provides a mechanism for returning to the question for another attempt (i.e., specific instructions to touch the screen, or a touch area labeled "Try again").

Fig. 3-8. Touch to try again—tailored feedbacks.

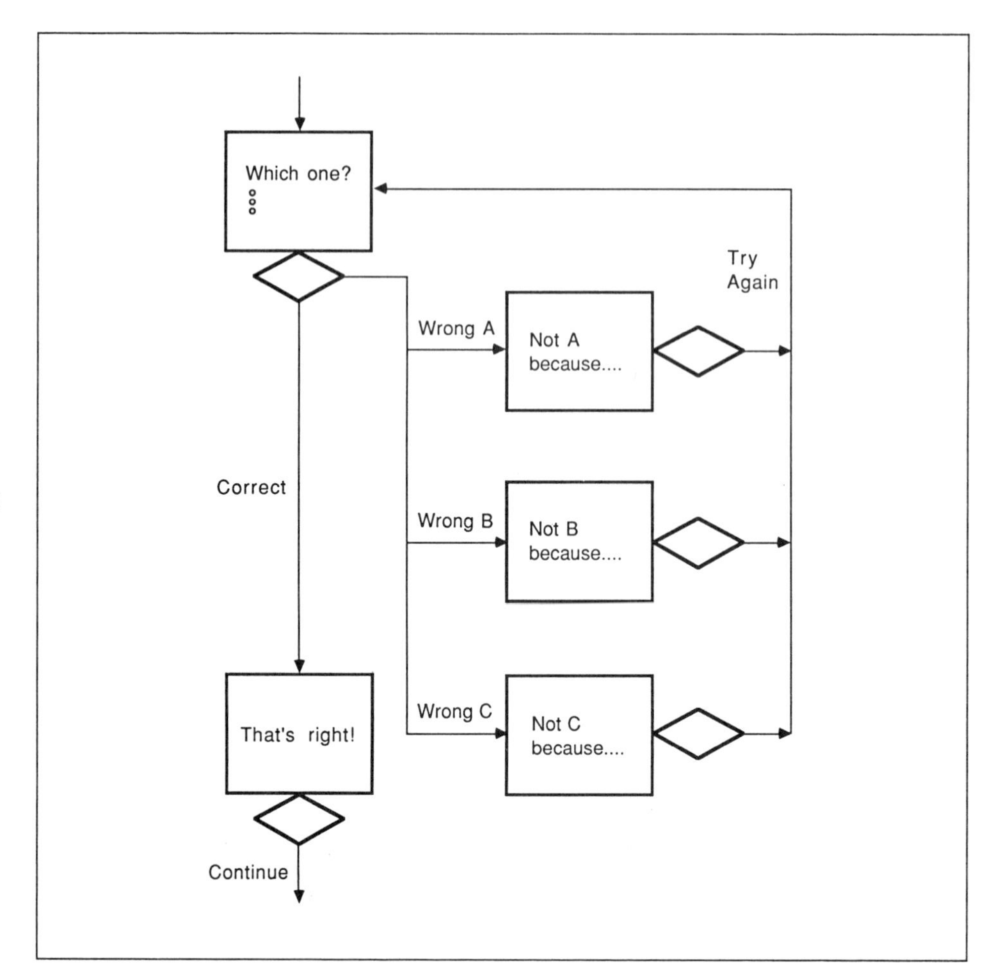

Advantage

Provides a mechanism for a repeated attempt at the question when the negative feedback destroys the original question or answers.

Disadvantages

Not particularly economical, since each incorrect response requires a specially written feedback.

Easily misused (chosen for situations that are inappropriate), creating unnecessary touches and *false touch options* (areas that appear to be touch active but are not).

Use When

Video takes up so much of the screen that the only way to provide feedback and still see the video is to overwrite (erase) the original question (and then only if it is not possible to provide the feedback and a restated question).

The feedback for the incorrect responses consists of a different video (in which case, the question should be erased to prevent visual confusion).

The feedback for the incorrect responses consists of a dramatic video and/or graphic response that erases the question.

There are three or four possible responses and they are obvious (so the student can get to the correct answer, if only by process of elimination).

Do Not Use When

Only two answers are possible and each one is obvious (the "Try again" strategy is not appropriate).

Feedback can be written on the same screen as the original question and answers, without altering the original video or graphics.

Development cost is a design issue, and the need for tailored feedbacks is not overwhelming.

The possible responses are not obvious (particularly when responses are video objects in a field of many; the student could be forever touching and still not hit the "magic" touch area; see Fig. 3-6).

It is possible to restate the question as part of the feedback, instead of erasing the question because of lack of room on the screen.

Feedback Overlaid—Generic Negative Feedback

Fig. 3-9 is similar to Fig. 3-5, except that the feedback has content and does not erase itself (does not time out). It is also similar to Fig. 3-7, except that a touch is not required to "Try again."

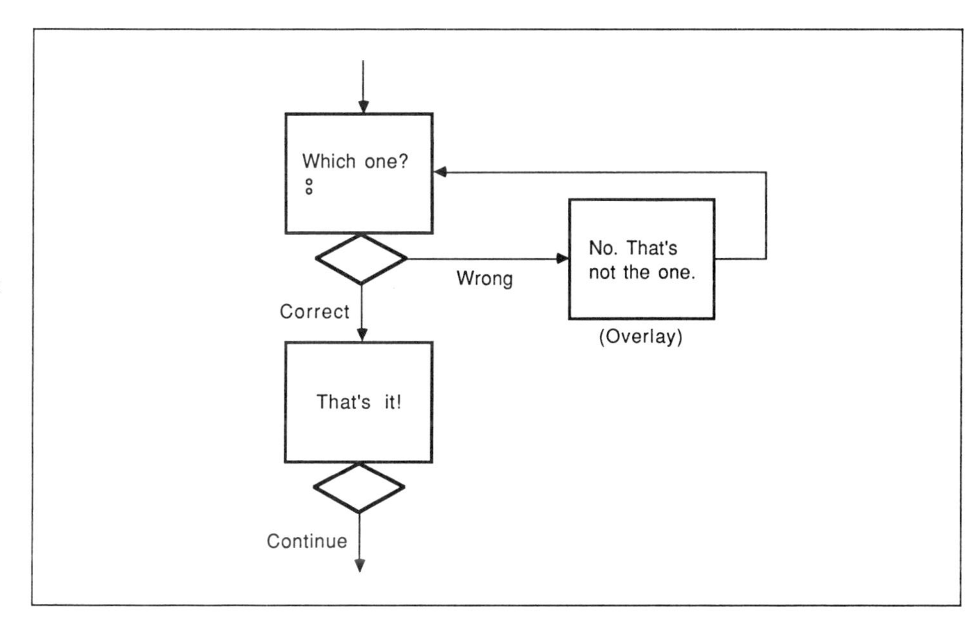

Fig. 3-9. Feedback overlaid—generic negative feedback.

Characteristics

Selecting the correct response leads to positive feedback.

Any incorrect response leads to a generic negative feedback, which is overlaid on the original question.

After an incorrect response, all original response touch areas remain active and the student can immediately make another attempt at answering the question.

Advantages

Effective visually when there is sufficient room for question, answers, and feedback all on the same screen.

Does not create useless touches. Every touch is an attempt to answer the question.

Does not leave false touch options on the screen (areas that appear to be touch active but are not).

Economical to develop because it requires only the question and two feedbacks.

Disadvantages

After the first incorrect response, further incorrect responses merely overlay the same feedback. If the student does not notice the feedback *blinking* (erasing and redrawing), he or she may assume that the computer is not responding.

It is sometimes difficult to write a single, generic negative feedback that fits in context with each of the possible wrong answers.

Can be frustrating to keep getting the same message for several wrong answers in a row. Not recommended for remedial students.

Creates the possibility of an endless loop if the student never gets the right answer.

Use When

There are four or fewer possible answers and they are obvious (the student can find the correct answer within a reasonable number of attempts).

Tailored negative feedbacks are not important.

It is instructionally important for the student to discover the answer independently, without the answer being given.

Do Not Use When

There are more than four possible answers (can be frustrating to keep getting the same message over and over).

Only two answers are possible and each one is obvious (the "Try again" strategy is not appropriate).

The possible responses are not obvious (particularly when responses are video objects in a field of many; the student could be touching forever and still not hit the magic touch area).

No change in the screen is perceptible when the feedback is overlaid on top of itself for the second and following wrong responses.

Creating a separate, germane negative feedback for each wrong response is easier than creating a single, negative feedback that makes sense for every wrong response.

The screen will not accommodate the question, the answers, and a negative feedback all at the same time.

Feedback Overlaid—Tailored Negative Feedbacks

Fig. 3-10 is similar to Fig. 3-9, except that separate feedbacks are created for each incorrect response. It is also similar to Fig. 3-8, except that no touch is required to try again.

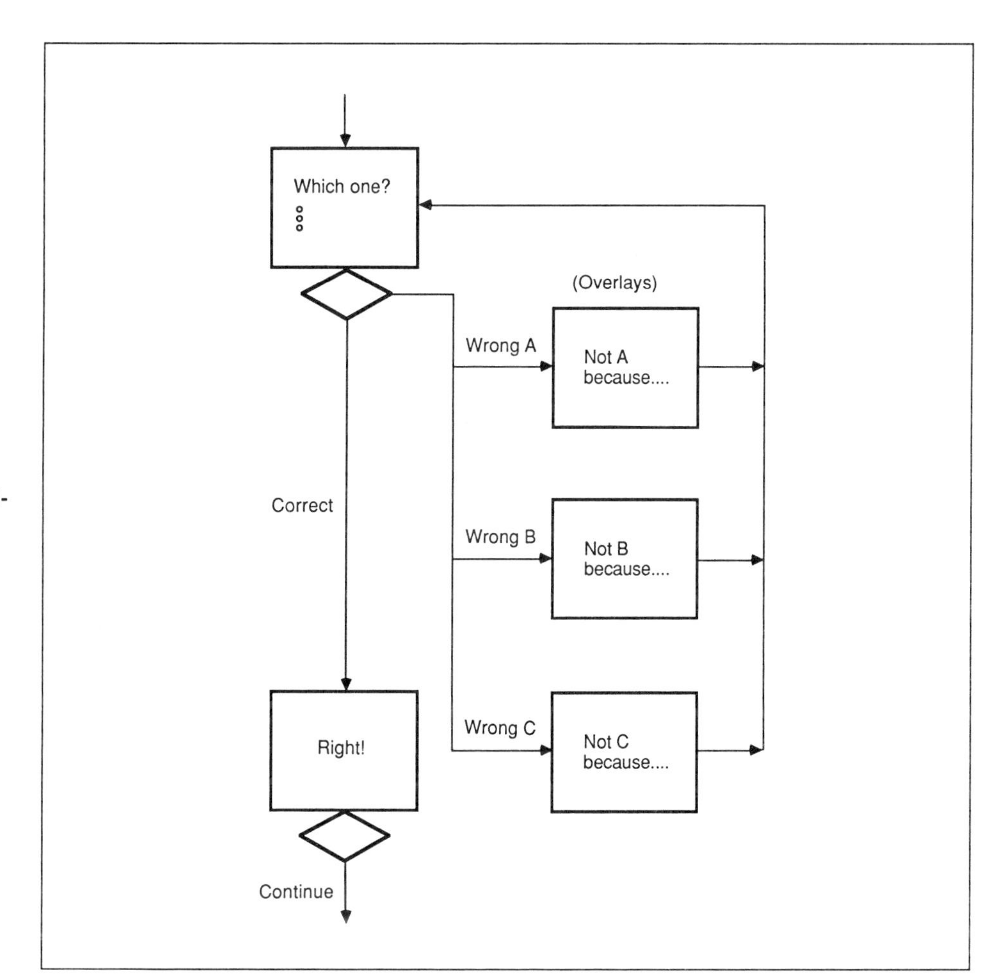

Fig. 3-10. Feedback overlaid—tailored negative feedbacks.

Characteristics

Selecting the correct response leads to positive feedback.

Each incorrect response leads to a unique negative feedback, which is overlaid on the original question.

After an incorrect response, all original response touch areas remain active and the student can immediately make another attempt at answering the question.

Advantages

Effective visually when there is sufficient room for question, answers, and feedback all on the same screen.

Does not create useless touches. Every touch is an attempt to answer the question.

Does not leave false touch options on the screen.

Effective instructional strategy for discovery learning, since every response gives immediate contextual feedback.

Disadvantages

Creates the possibility of an endless loop if the student never gets the right answer. Not recommended for remedial students.

Requires additional effort to write individual feedbacks for each incorrect response.

Use When

There are three to six possible answers (at most eight) and they are obvious (the student can find the correct answer within a reasonable number of attempts). The "reasonable number of attempts" is higher for this interaction type because of the variety and constructive nature of tailored feedbacks, and the immediate nature of feedback and reattempt.

It is instructionally important for the student to discover the answer independently, without it being given.

Do Not Use When

There are more than eight possible answers (it may be frustrating for the student to try seven or eight times and still not get the correct answer).

Only two answers are possible and each one is obvious (the "Try again" strategy is not appropriate).

The possible responses are not obvious (particularly when responses are video objects in a field of many; the student could be touching forever and still not hit the magic touch area; see Fig. 3-11).

Development cost is a design issue and there is no overwhelming need for tailored feedbacks.

The screen will not accommodate the question, the answers, and a negative feedback all at the same time.

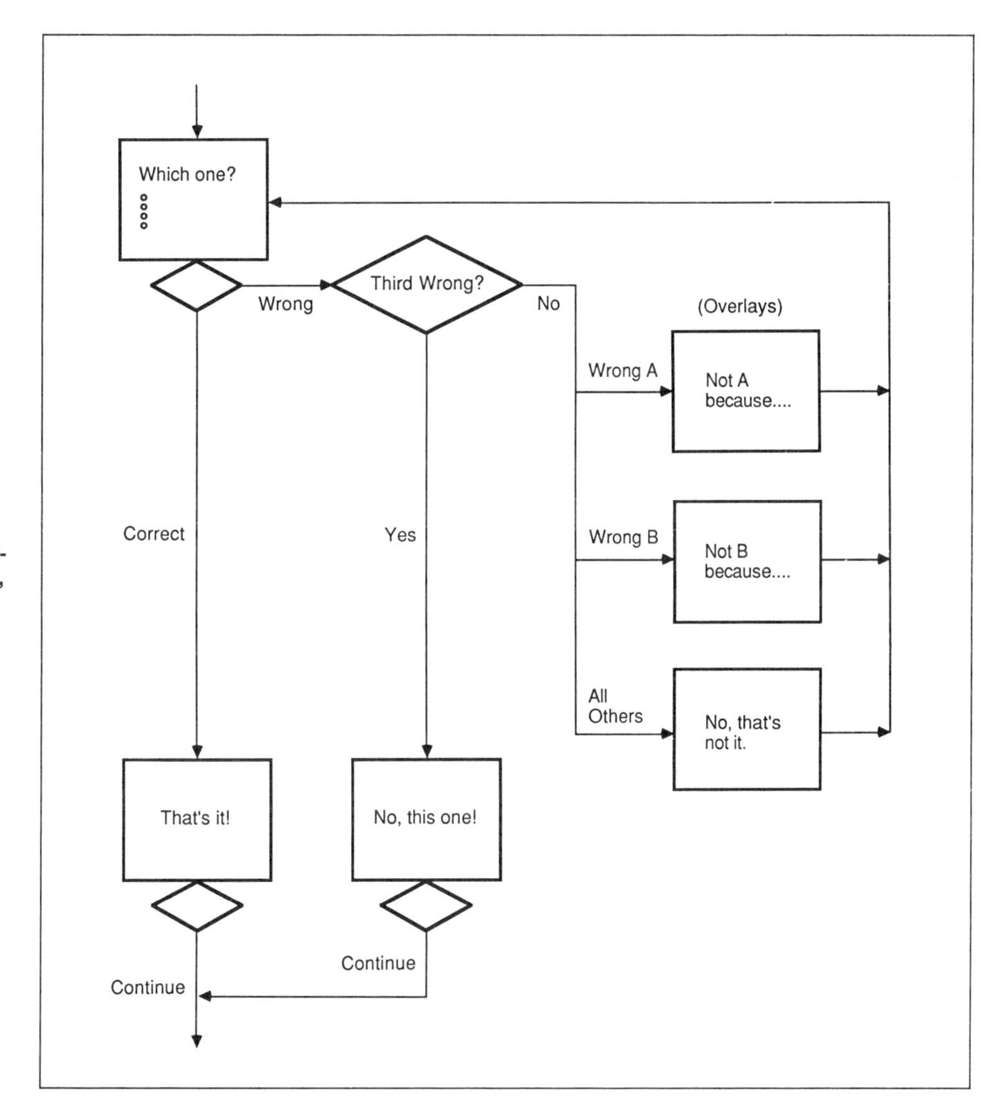

Fig. 3-11. Feedback overlaid—tailored and generic, with counter.

Feedback Overlaid—Tailored and Generic, with Counter

Fig. 3-11 is a variation of Fig. 3-10. It illustrates the use of a counter and combines the tailored and generic feedbacks. Feedbacks A and B are specific to incorrect responses A and B. The third feedback is generic and applies to any other incorrect responses (assuming that four or more wrong responses are possible).

This synthesis of interaction styles capitalizes on the advantages of generic and tailored feedbacks, and eliminates the weaknesses associated with endless looping. This is one of the most versatile and effective interaction types.

Characteristics

Selecting the correct answer leads to positive feedback.

The first few incorrect responses give feedback overlaid on the same screen with the question and answers, and the student has an opportunity to answer the question again.

A uniquely worded feedback is provided for certain incorrect responses. All other incorrect responses lead to an identically worded, generic feedback.

After a predetermined number of incorrect responses, the student is given the answer.

By definition, this strategy can be used only with questions having at least four possible answers (one correct, one with specific feedback, two with generic feedback).

Advantages

Allows for tailored feedback when it is appropriate.

Does not force tailored feedback when inappropriate.

Eliminates the problem of the endless loop.

Disadvantages

Not particularly economical because of the variety of feedbacks that must be provided and the variety of branching options.

Screen may appear to be dead when two incorrect responses in a row lead to the same generic feedback (if it is not apparent that the feedback is erasing and redrawing).

Use When

Instructional effectiveness is a priority concern over cost.

The question, answers, and feedbacks will all fit on one screen.

Do Not Use When

Cost is a primary concern and another, less exotic, interaction will meet instructional needs.

The question, answers, and feedbacks will not all fit on one screen.

Feedback Overlaid—Generic Negative Feedback, with Counter

Similar to Fig. 3-9, the strategy shown in Fig. 3-12 incorporates a counter to eliminate the problem of endless looping. It can also be viewed as a variation on Fig. 3-3, where the student is given a limited number of retries before being given the answer. It may also be viewed as a variation on Fig. 3-6, except that the feedback has content and does not erase itself (time out).

Characteristics

Selecting the correct answer leads to positive feedback.

The first few incorrect responses give a generic feedback overlaid on the same screen with the question and answers, and the student has an opportunity to answer the question again.

After a predetermined number of incorrect responses, the student is given the answer.

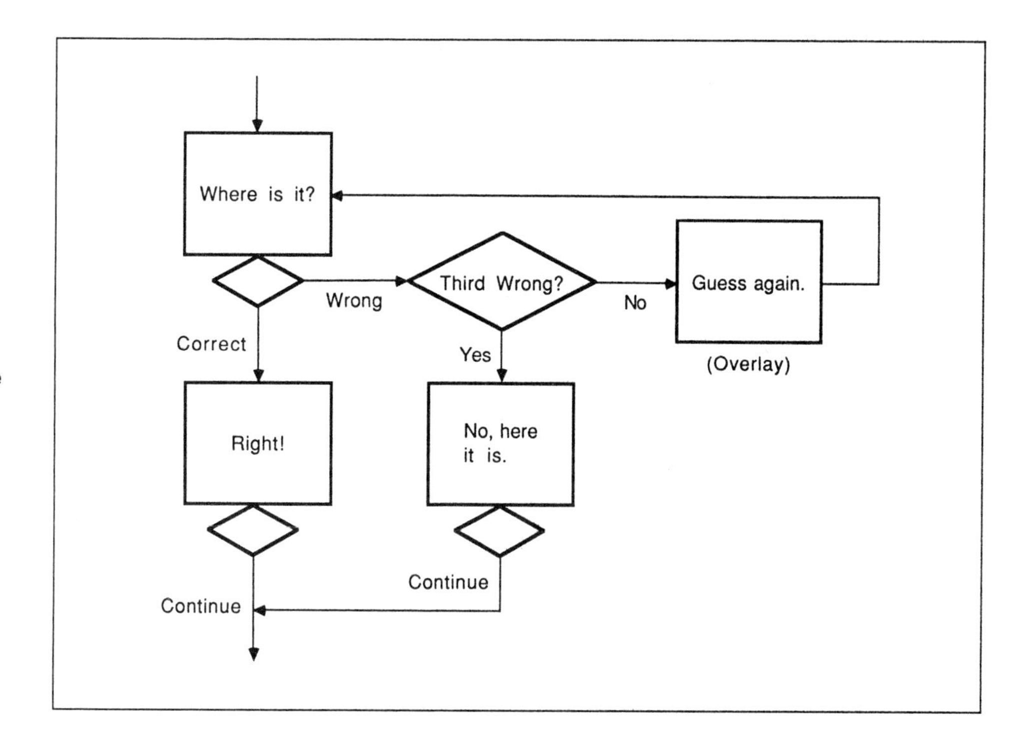

Fig. 3-12. Feedback overlaid—generic negative feedback, with counter.

Advantages

Does not require wording of feedbacks for every incorrect response.

Prevents the problem of an endless loop.

Disadvantages

Does not provide tailored feedbacks.

A generic feedback that fits all contexts may be difficult to create.

Use When

A counter is important for reducing frustration, or because possible responses are video based and not obvious.

Tailored feedbacks are not appropriate or desirable.

The question, answers, and feedback will all fit on one screen.

Do Not Use When

It is easier to create tailored feedbacks than it is to create a single, generic feedback that fits all contexts.

It is important to identify and dispel misconceptions that may be associated with particular incorrect responses (tailored feedbacks are needed).

The question, answers, and feedbacks will not fit on the same screen.

Only two answers are possible and each one is obvious (the "Try again" strategy is not appropriate).

Give Answer after Second Wrong

Fig. 3-13 shows what might be considered a special-case version of Fig. 3-12. When the wrong-answer counter is set for two wrong answers, it may be just as economical (and possibly more flexible) to use a second question frame instead of a counter. The second-chance question frame is the same interaction type as that in Fig. 3-3.

Characteristics

Selecting the correct response leads to positive feedback.

The first wrong response provides some form of negative feedback, hint, or restatement of the question and allows a second attempt.

The correct response on the second attempt leads to the same positive feedback.

An incorrect response on the second attempt leads to the answer being given.

Advantages

Does not require a counter, although it behaves as if it has a wrong-answer counter set for two wrong answers.

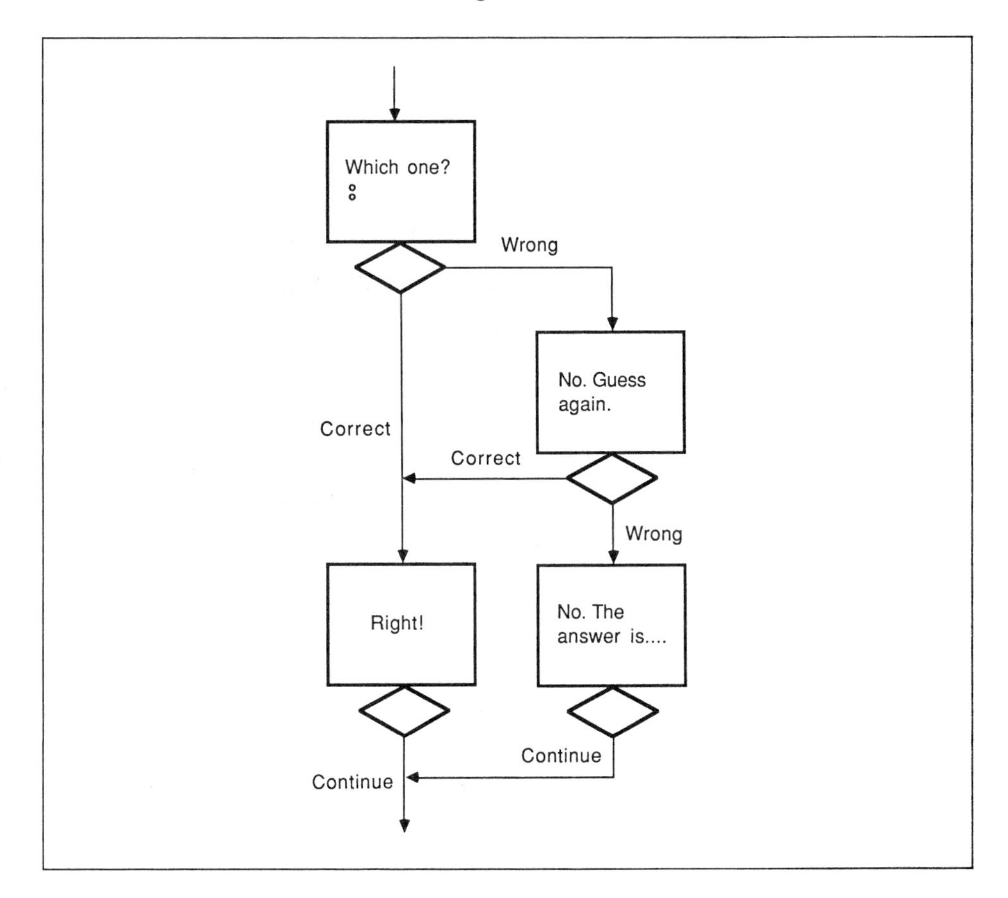

Fig. 3-13. Give answer after second wrong.

Branching forward on an incorrect response, instead of backward, allows subtle changes in the wording of the question, the use of additional video, or other changes in the question/answer presentation that may help the student get the answer right.

Minimizes frustration while giving the student a second chance to get the answer right.

Economical and effective strategy requiring only four elements:

1. Original question.
2. Correct feedback.
3. First wrong answer feedback.
4. Second wrong answer feedback.

Disadvantage

Does not provide specific feedbacks for each incorrect response (feedbacks not tailored).

Use When

It is important for the student to have a second attempt at a question.

A simple strategy that is effective, flexible, and economical is needed.

Do Not Use When

Only two answers are possible and each one is obvious (the "Try again" strategy is not appropriate).

It is important to identify and dispel misconceptions that may be associated with particular incorrect responses (tailored feedbacks are needed).

Give Answer after Third Wrong

Fig. 3-14 is a variation of Fig. 3-13, since it provides one more opportunity to get the answer right before it is given. This strategy might be compared to a wrong-answer counter set at 3, but there is a significant conceptual difference. The feedbacks vary, not with the particular wrong answer chosen, but with the *number* of wrong answers chosen.

Characteristics

Consists of three levels of the same question. Each time the question is missed, the student advances to the next level, where the question is presented again with feedback, altered wording, or hints.

Selecting the correct response at any level leads to the positive feedback.

When the third-level phrasing of the question is missed, the answer is given.

By definition, the question must have at least four possible answers.

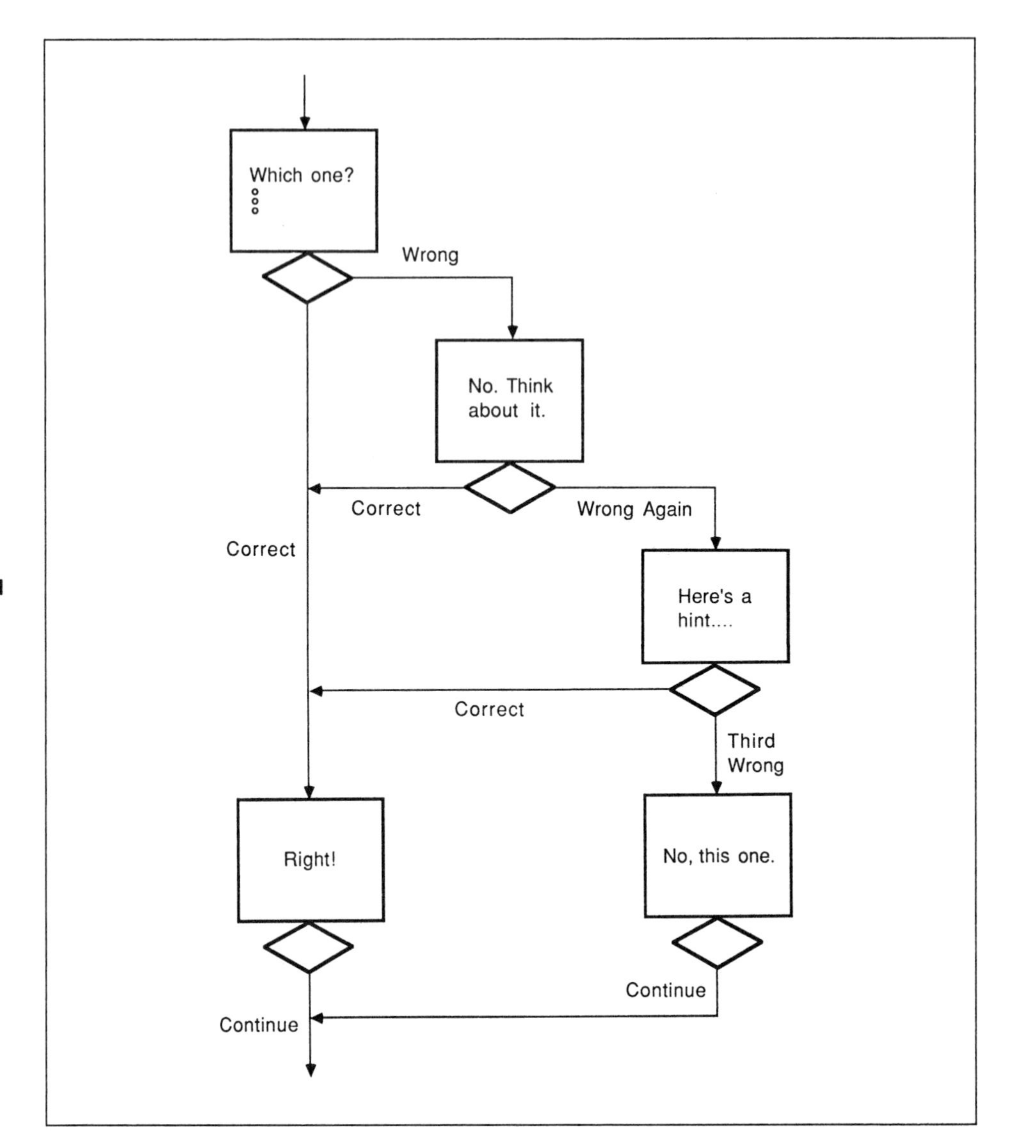

Fig. 3-14. Give answer after third wrong.

Advantages

Branching forward provides greater flexibility in rephrasing a question that a student is finding troublesome. The question can be rephrased, the video made more explicit, or hints given as appropriate.

Can address the needs of a wide variety of students. Two follow-up versions of the question allow for a gentle hint when the question is missed once, and a stronger hint when the question is missed twice.

Disadvantages

Does not provide specific feedbacks for individual, incorrect responses (no tailored feedbacks).

Sometimes difficult to come up with three different ways of asking the same question, or with appropriate levels of feedback and hints for first and second incorrect answers.

Use When

The question presentation is difficult and may not be understood by all students. This interaction type allows the designer to take a chance with the first question but does not penalize the student for not understanding on the first try. The question can be rephrased and asked again up to two more times.

Do Not Use When

It is important to identify and dispel misconceptions that may be associated with particular incorrect responses (tailored feedbacks are needed).

Feedback of Mixed Types

There is no rule that all feedbacks have to be the same type, as long as each is used appropriately. Fig. 3-15 illustrates the use of an overlaid feedback, a timed-out message, and a "Try again" button.

ADDING AUDIO AND MOTION

The previous examples illustrated the use of video still-frames (with some motion) combined with static graphics. Although these strategies can be used to provide simulation of cause and effect in certain applications (touch a switch, the switch flips off, and an indicator light comes on), there are many situations in which the result of a student's choice can be most effectively illustrated by using motion. And there are some situations to which the student must respond that can only be presented realistically with motion.

Another benefit of adding motion sequences in place of still-frames is the ability to incorporate audio questions and feedbacks, even if the video does not change. This benefit could also be achieved using various sound-over-still technologies.

If sound-over-still is used or if video motion is used exclusively with first or second audio track, with no perceived change in the video, then all of the previous figures can be used without alteration.

When adding audio elements to the interactions previously described, keep the following question in mind: Can the student re-create the audio question or feedback if he or she missed hearing or understanding it the first time? The reality of a learning environment is that interruptions and distractions do occur.

One of the criteria in designing interactions should be measured against the "walkaway" test. If the student walks away from the system and returns later, does the screen present enough information that the student can recall specifics, and does it provide an opportunity for the student to re-create the situation? Student frustrations could arise if audio is played and no opportunity is provided to hear the question again.

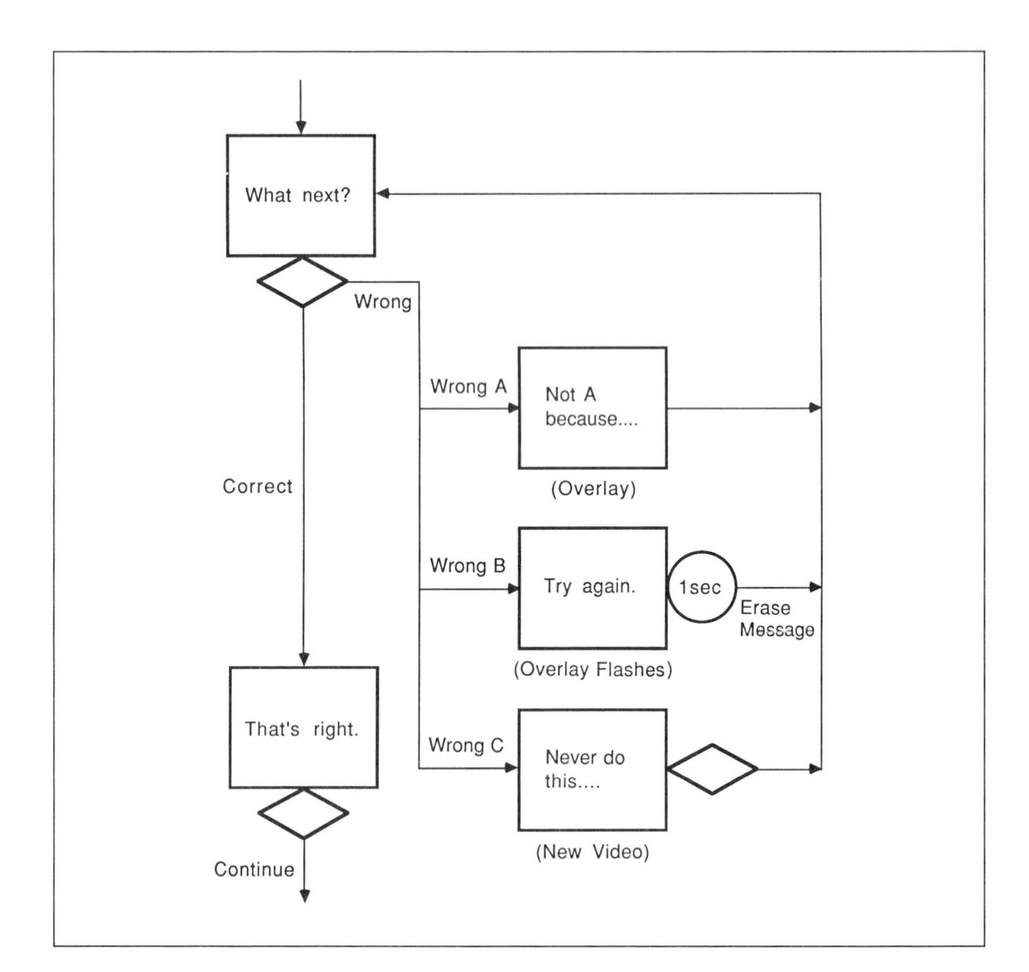

Fig. 3-15. Feedback of mixed types.

If audio feedback is used for a negative response, the consequences of the student not hearing it may not be severe. If, however, the student's response to a situation poses disastrous consequences, something more enduring than an audio response should be provided—either a dramatic visual response or a text response giving the student an opportunity to study the described result of his or her action.

Using Motion to Create the Question Situation

In most of the previous figures, a motion sequence can be substituted for video stills if certain factors are taken into account.

Any question posed to the user can be set up by using a motion sequence that freezes on the question situation. In Figs. 3-1 through 3-4 and Figs. 3-13 and 3-14, the original question is posed only once and the substitution of a motion lead-in is straightforward. Fig. 3-16 illustrates a variation of Fig. 3-4, using a motion lead-in and a question replay option.

When adding a video motion element as a lead-in to any of the questions in Figs. 3-5 through 3-12, the designer must take into consideration those situations in which a feedback alters the video content of the question frame. The

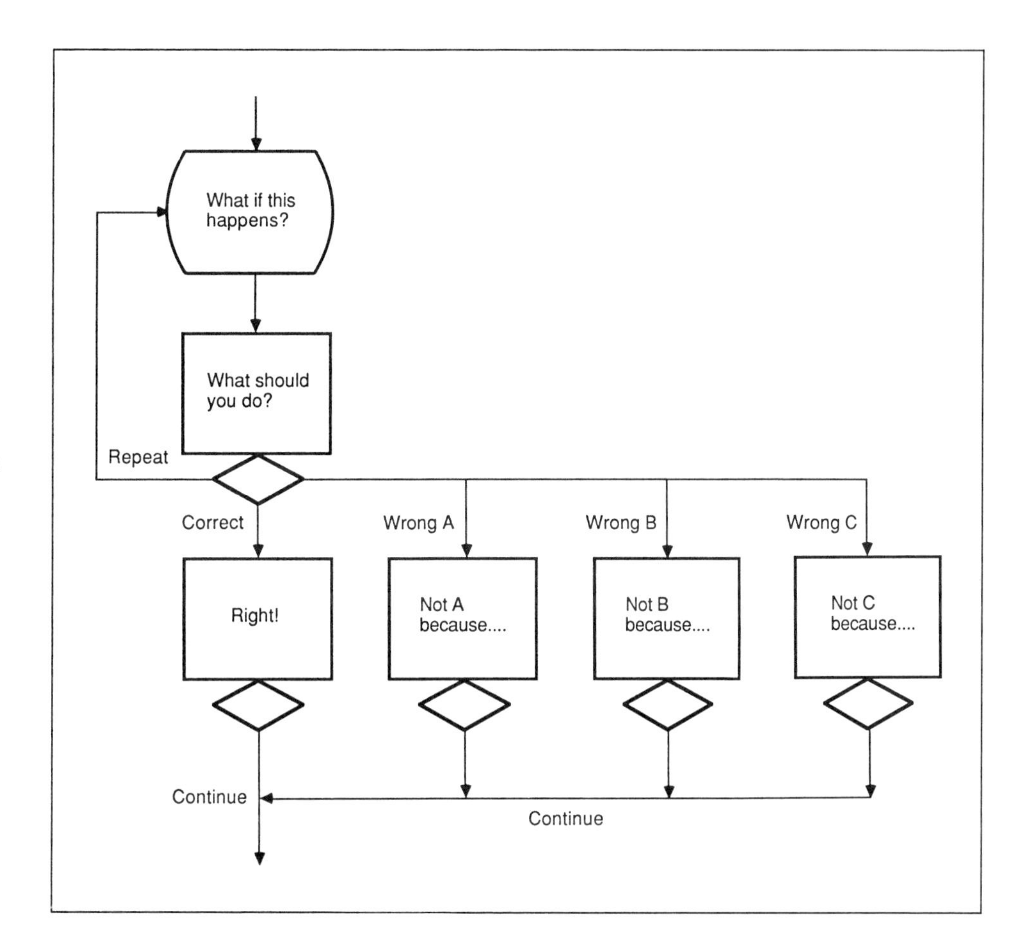

Fig. 3-16. Motion lead-in to question.

designer must decide whether to replay all or part of the motion lead-in or to merely return to the freeze frame that posed the question.

Fig. 3-17 illustrates a motion lead-in that has been logically divided into two parts to provide a unique entry point for a restatement of the question. The feedbacks illustrated are of two types: two that do not alter the video (wrong A and wrong D) and two that do. In the context of this example, it is assumed that for wrong answers A and B it is not necessary to replay the motion lead-in, but merely to return to the question situation. For wrong answers C and D, however, it is assumed that the student misunderstood the question, and part of the question is replayed.

If the question itself is posed in audio as part of the motion lead-in, one should also consider the effect of replaying the same audio when the question is posed a second or subsequent time. If repeating the same audio is in danger of becoming boring or trite, a variation on Figs. 3-13 or 3-14 can be used.

Fig. 3-18 illustrates another variation of Fig. 3-13, which uses audio track 1 over motion for the first question situation and then uses audio track 2 over the same motion sequence if the question is repeated. Of course, an entirely new and unique motion sequence could be used for the second question situation.

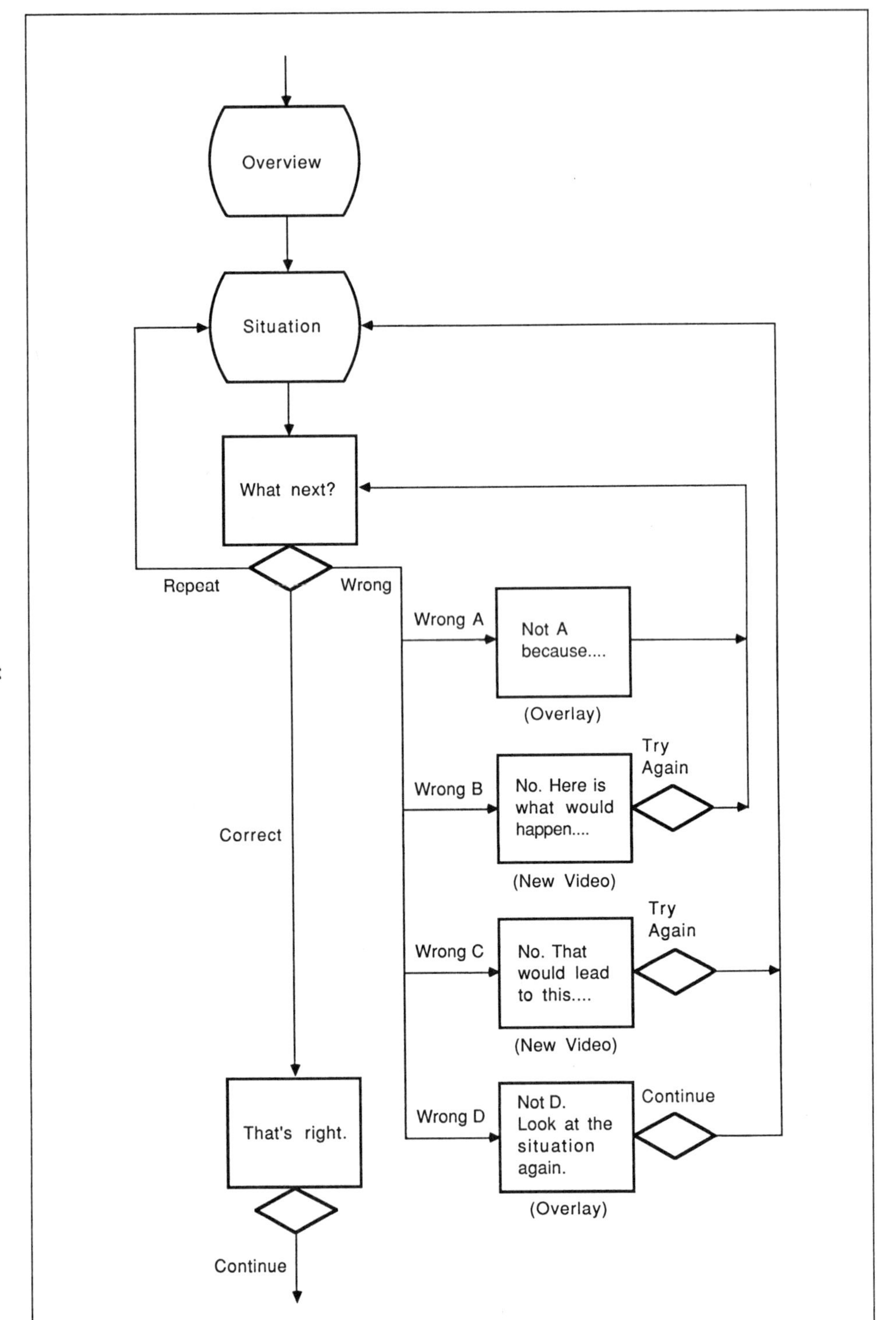

Fig. 3-17. Two-part motion lead-in.

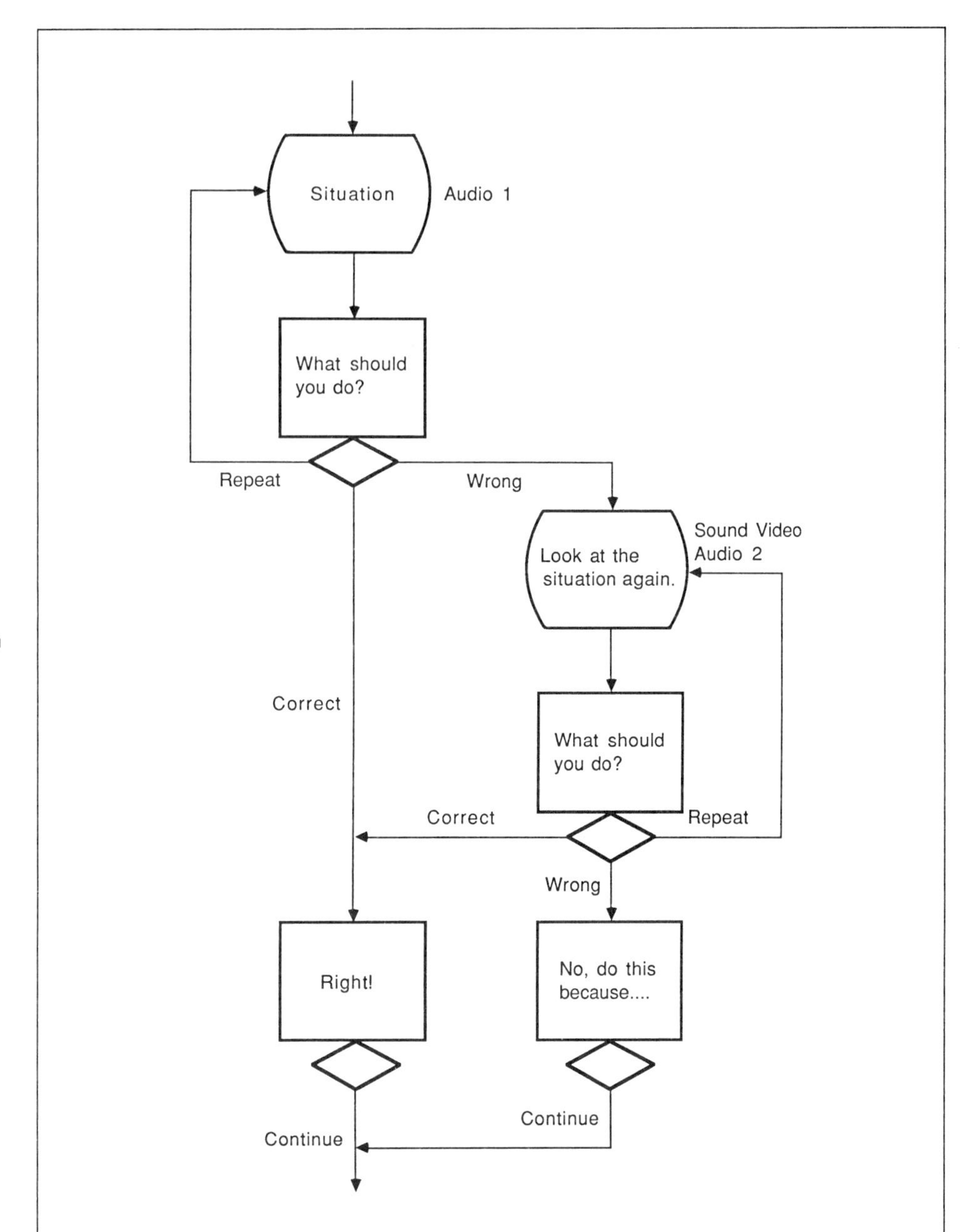

Fig. 3-18. Dual audio for question restatement.

Using Motion to Provide Positive and Negative Feedbacks

In any of Figs. 3-1 through 3-15, a motion sequence can be used as a feedback. When a motion response is used to give negative feedback to a student action, the video component of the original question frame is obviously altered. If the student is to be given another chance to respond to the question, the original situation must be re-created—either by returning to the original still-frame or by replaying all or part of the motion lead-in to the question.

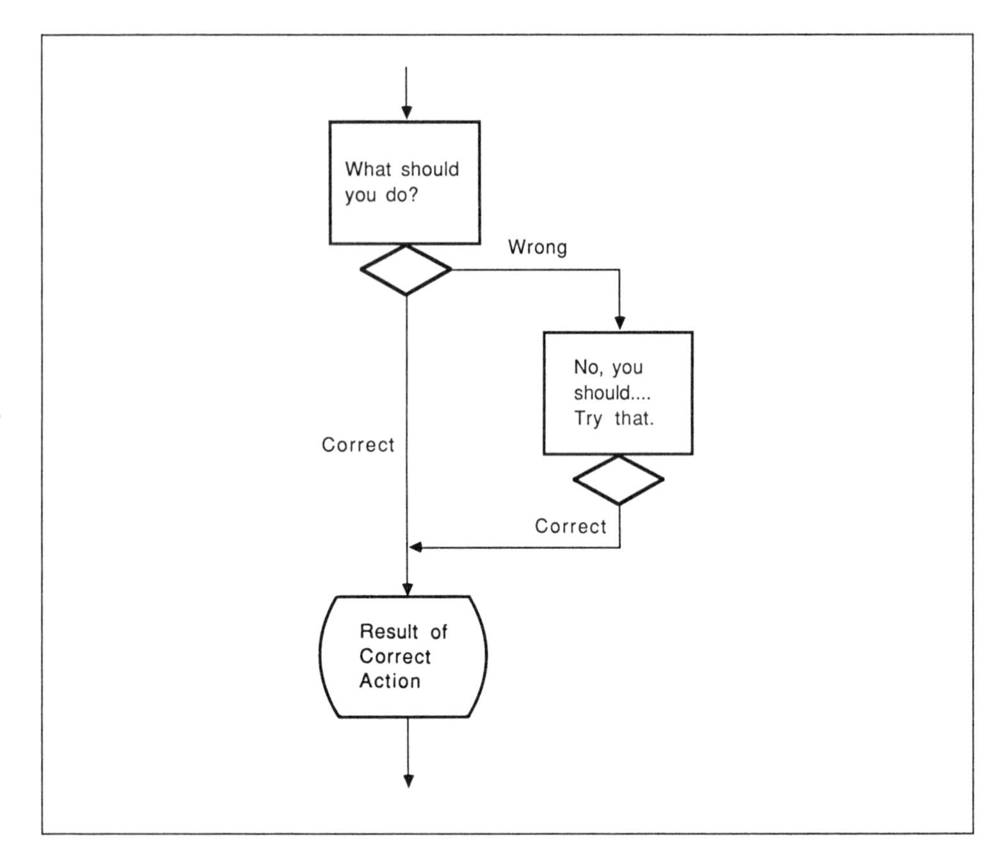

Fig. 3-19. Positive feedback using motion.

Figs. 3-19 through 3-23 illustrate specific variations of previous interactions that incorporate motion feedbacks. These figures are by no means exhaustive. Each of the various feedback strategies should be measured in terms of coherency and effectiveness; that is, does the strategy make sense when presented in that sequence and does it provide effective learning?

SIMULATIONS

In real life, time seldom stops while we wait to make up our minds about taking a particular action. In all of the previous figures, the program stops while the student decides what action to take. Moreover, real-life choices are seldom conveniently contained within a single video scene. In short, simulating real life with interactive video requires more than has been illustrated so far.

To create simulations, two unique varieties of interaction can be added to the repertoire: *position-sensitive* and *time-sensitive* interactions.

In a position-sensitive interaction, the student must arrive at some predetermined position before taking action. This may involve adjusting a knob to attain a proper meter reading, or moving physically around a room or building. The student's "position" is equated to a videodisc frame. The screen presents one or more action options and some facility for changing position, which results in a change in the video.

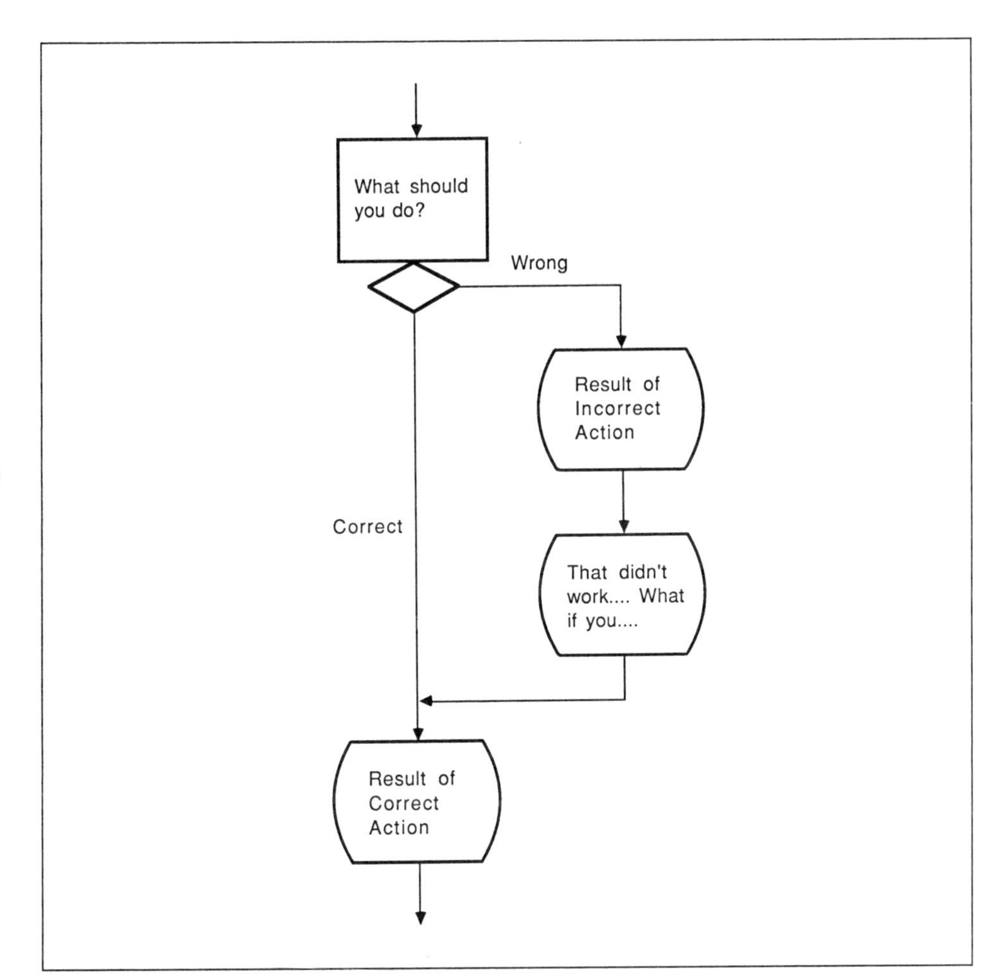

Fig. 3-20. Positive and negative feedback using motion.

Fig. 3-24 illustrates a simulation for setting a pressure valve and then toggling a switch when the pressure is correct. The screen illustrates a pressure meter, a control knob, and a switch. The student's options are to lower the pressure, increase the pressure, or toggle the switch. Once the switch is toggled, three outcomes are possible, depending on the pressure (that is, the position of the needle on the meter). If the pressure is increased beyond the limit, a fourth outcome is possible.

The variety of uses for the position-sensitive interaction in gaming and strategy applications are virtually limitless. Another form of this type of interaction could be used to illustrate proper procedures for lifting a heavy object, as illustrated in Fig. 3-25. The student steps through various still-frames depicting someone about to pick up a large box and then chooses to lift the box from that position. Only one of the stills illustrates the proper lifting posture. If the student elects to lift and is not in the proper position . . . ouch!

The time-sensitive type of interaction addresses real-time simulations, in which an action is appropriate only at specific times. This can be viewed as a variation of the position-sensitive interaction, except that the student is no longer in control of the position. The position is changing in real time, and the student must identify the correct position (or time) to take action as it occurs.

Fig. 3-21. "Try again," with generic negative feedback using motion.

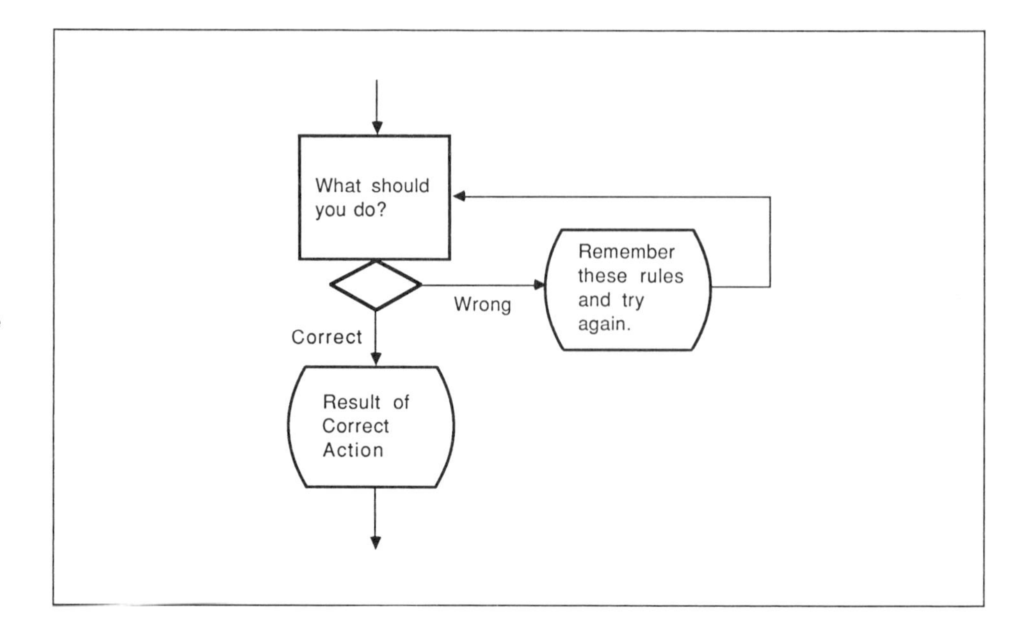

Fig. 3-22. Tailored feedbacks using motion.

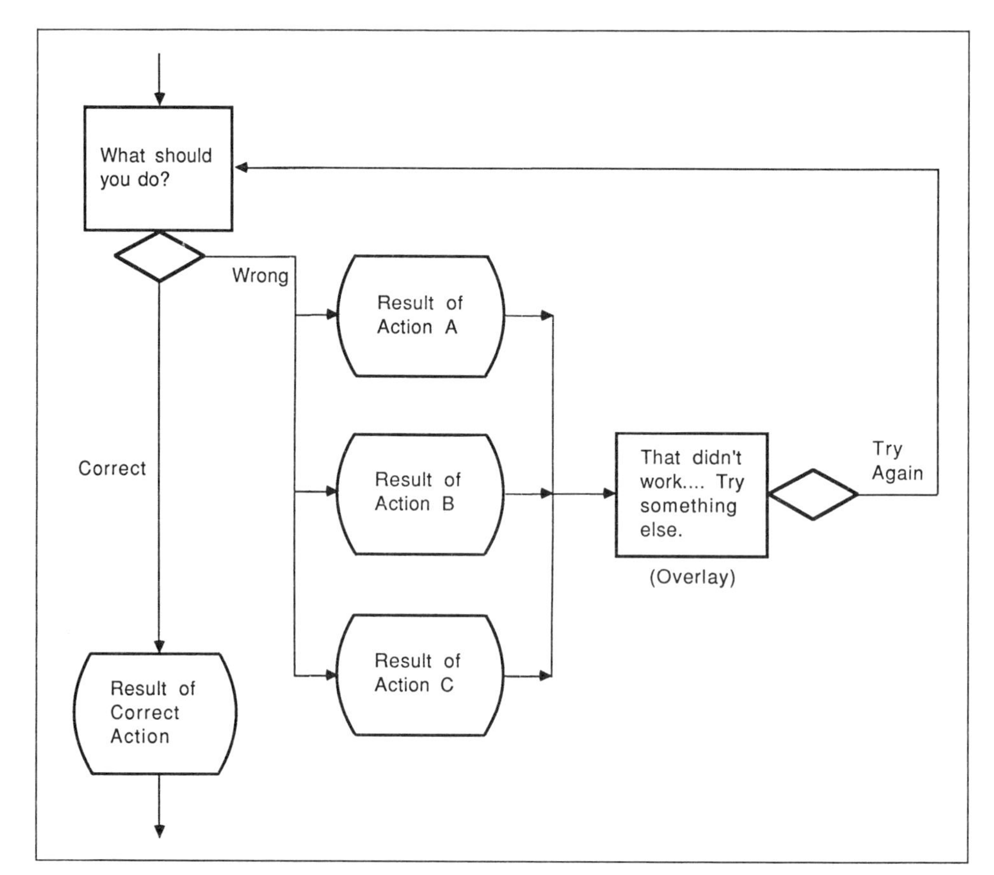

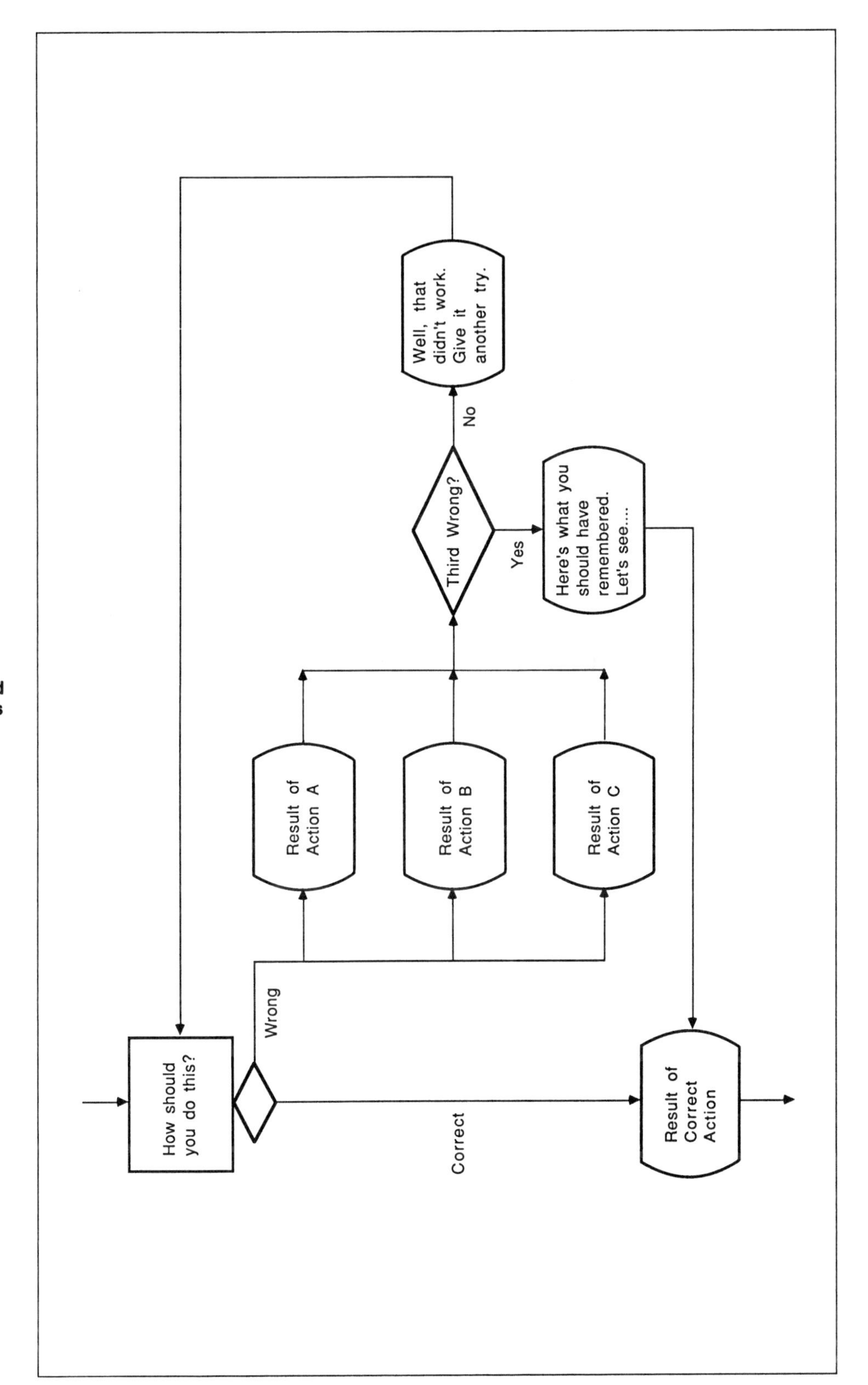

Fig. 3-23. Tailored motion feedbacks with counter.

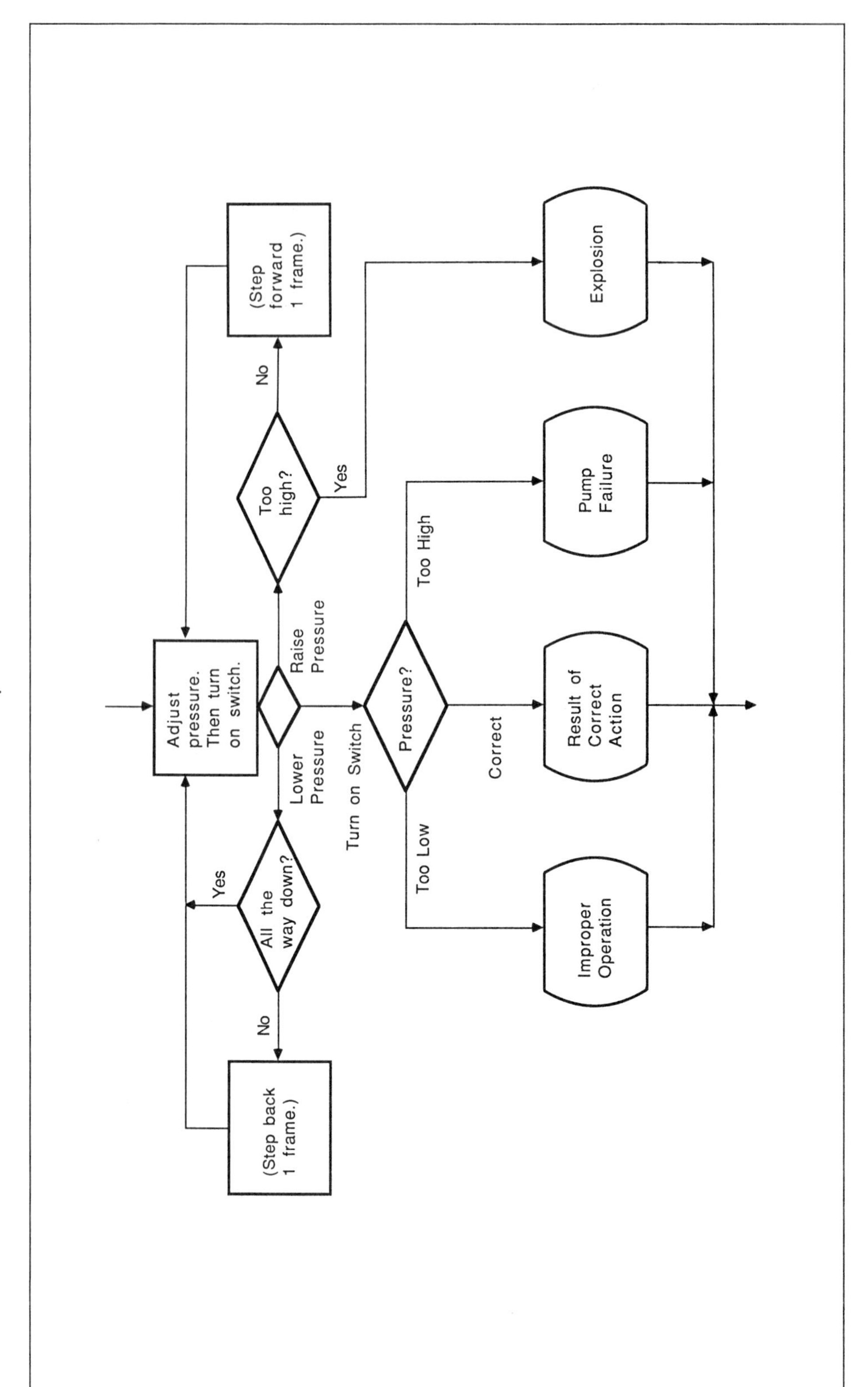

Fig. 3-24. Position-sensitive simulation.

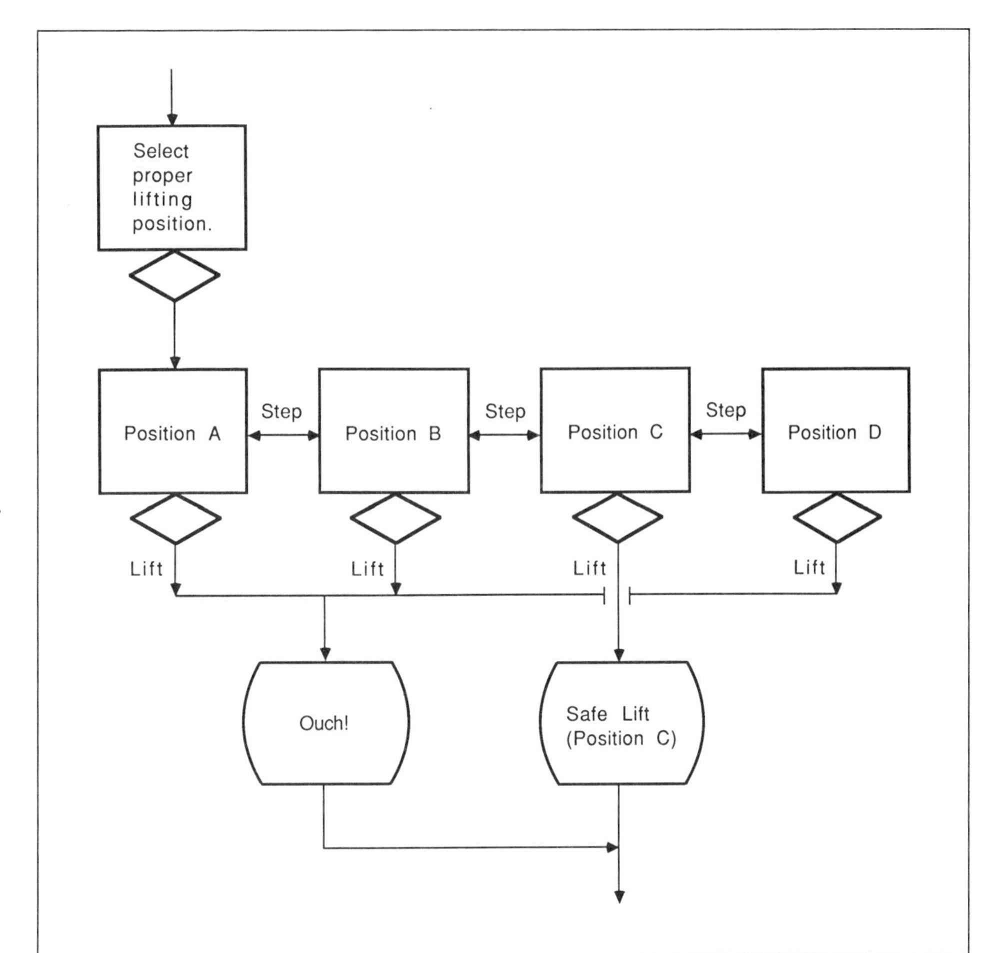

Fig. 3-25. Another example of position-sensitive simulation.

Fig. 3-26 illustrates a simulated refueling operation, in which the student has been assigned to stand by for emergency fuel shutoff. He or she has previously received instructions on the conditions for using the emergency shutoff. As the scene unfolds, various subtle events begin to occur. The most critical is a colonel approaching with a lit cigarette. According to the student's training, it would be appropriate to activate the emergency shutoff if the colonel comes within 100 feet carrying the ignition source. If the student reacts too soon, the video scene changes to show the colonel extinguishing the cigarette because he or she is quite cognizant of the 100-foot limitation, and then becoming irate because the refueling has been halted. If the colonel gets within 50 feet, an explosion and fire result. Between 100 and 90 feet, the student is congratulated for prompt and appropriate action. Between 90 and 50 feet, the student is both rewarded for appropriate action and reprimanded for waiting so long to react.

COMBINING INTERACTIONS

Interactions are merely building blocks to be used in constructing an interactive program. With a solid understanding of basic interaction types and their appropriate uses, it is easy to create a program that provides effective and engaging

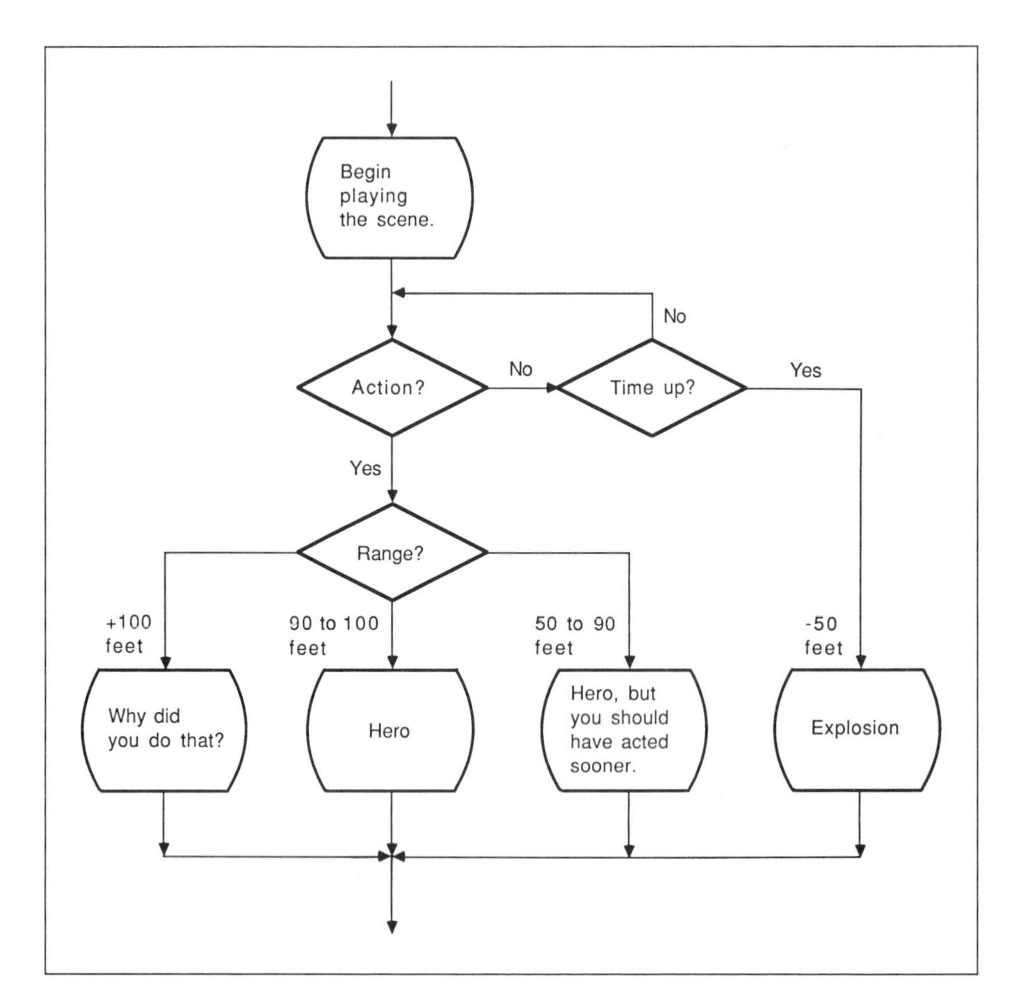

Fig. 3-26. Time-sensitive simulation.

instruction, that is economical to develop, and that is much more than an "electronic page turner."

Using the basic building blocks, the designer can avoid becoming entangled in complex branching schemes that sometimes result from ambitious attempts at simulation. One method for keeping the program under control is to select the appropriate interaction for the particular objective at hand.

The interactions described thus far have had one feature in common: Each has a single entry and exit point. This intentional feature enables a wide variety of interactions to be easily chained together, as illustrated in Fig. 3-27. This example illustrates testing a student's ability to perform a six-step procedure. Each interaction type is chosen for its appropriateness to the procedural step being tested. The result of chaining the basic interaction types is that remediation is simplified by being localized to each step, and the branching is extremely straightforward. It is linear, in fact, when viewed from the macro level (shown on the right).

Fig. 3-27 shows a micro view (left) and a macro view (right) of a linear combination of interactions. However, certain situations may require more than a linear assemblage of interactions. Using the basic building blocks, it is possible

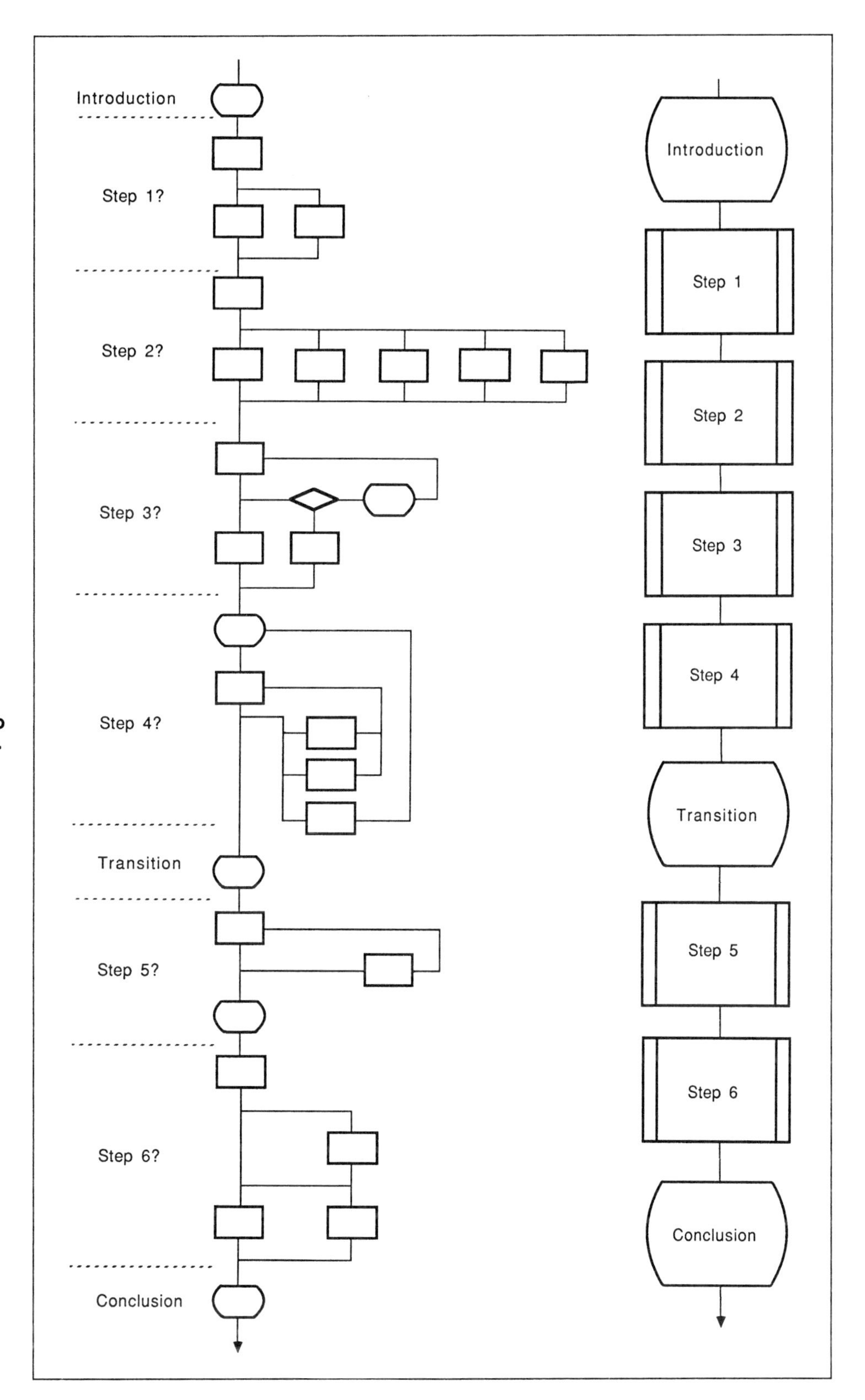

Fig. 3-27. Linear combination of interactions (micro and macro views).

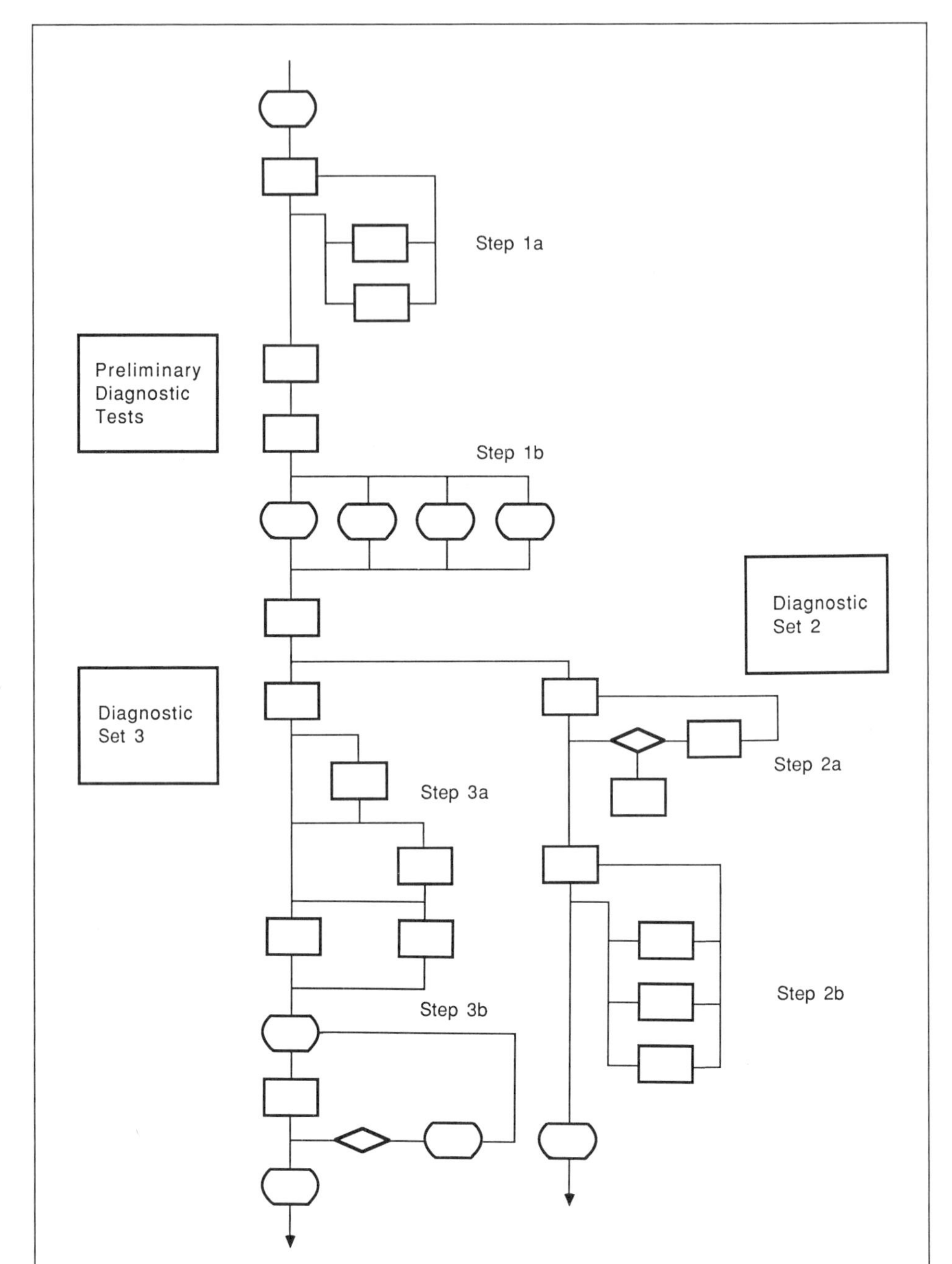

Fig. 3-28. Major branches in combined interactions.

to maintain control of the program by creating alternate chains of interactions as shown in Fig. 3-28. In this figure, the logical structure of the program has two major branches. After performing preliminary diagnostic tests, the student may choose to perform one or the other of two additional diagnostic procedures to determine what is wrong with a piece of equipment. Branching can be as complex as necessary; but, by using modular collections of interactions, the program and its development costs can be kept under control.

Fig. 3-29. Simulation involving the packing of a parachute.

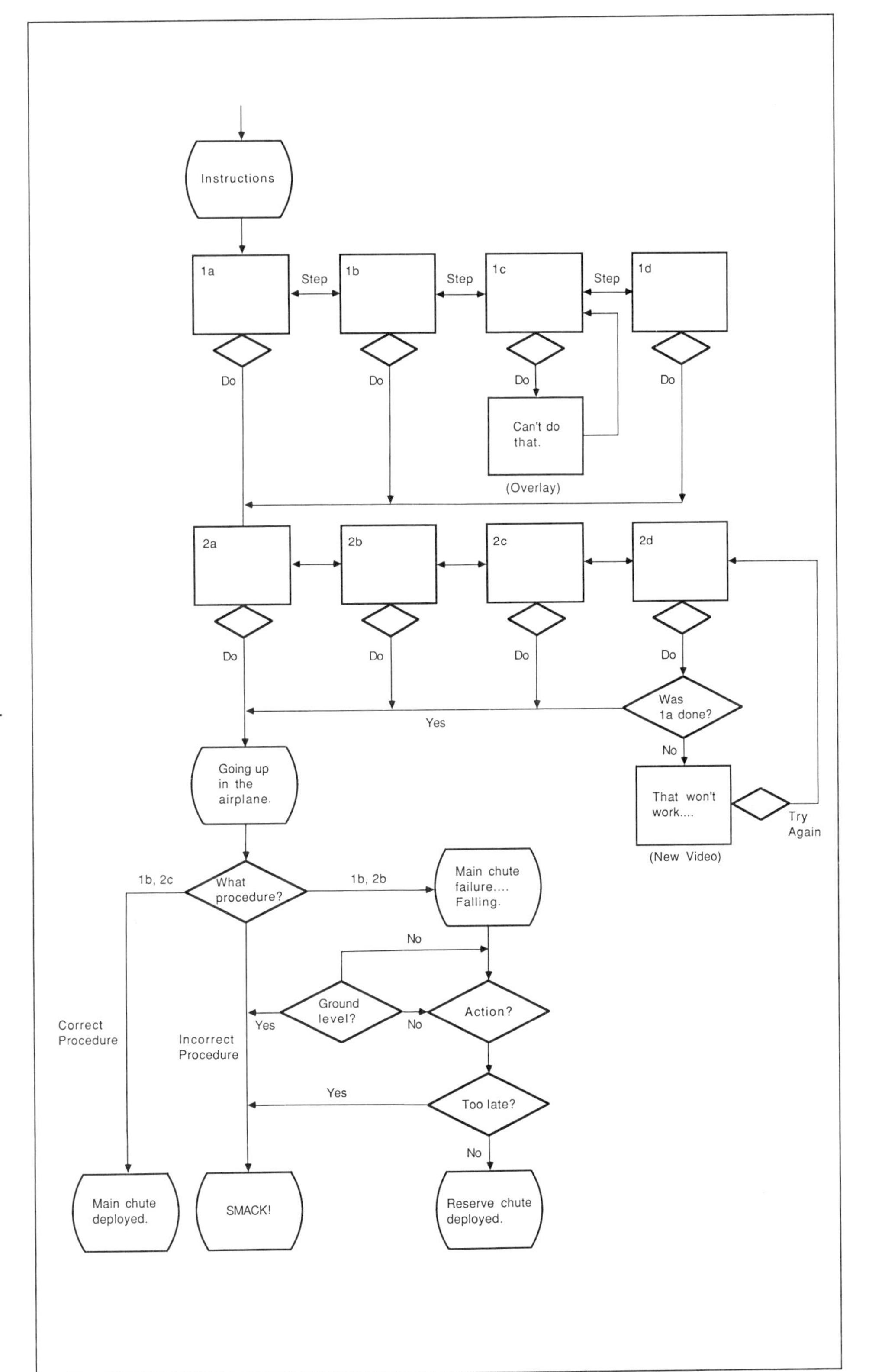

One final example illustrates the use of student history to control the program's branching. It is common for a student to be scored over a series of interactions and the net score to be used for controlling the student's subsequent path in the program. For example, if a student fails a test, the student is directed to a remedial learning component; but, if the student passes, he or she is directed to the next component.

In certain simulations, it is necessary to present several interactions in a series, with the final outcome dependent on choices made throughout the series. As the student progresses through each interaction, feedback may not be supplied when certain incorrect actions are taken. In fact, a given action may be incorrect only if a subsequent action is not also taken. At the end of the series, a history of the student's choices is analyzed to determine the outcome.

Fig. 3-29 is an overly simplified example of a simulation involving the packing of a parachute. This examples uses a series of positional interactions enabling the student to select pictures of appropriate procedures for packing the parachute. Certain selections are nonsensical and must be corrected before the simulation can logically proceed. Other actions are incorrect only if a previous or subsequent action is not also performed. Some of these situations are remediated when they occur; others are temporarily overlooked until the conclusion of the exercise.

At the end of the simulation, the student goes up in an airplane and jumps out. One of three outcomes is provided, depending on the student's history of choices. The third outcome includes a time-sensitive interaction requiring the student to activate his or her reserve parachute before reaching a critical altitude.

4

Level II Videodisc: Some Considerations for Programmers

Rick Kent, M.D.

Dr. Rick Kent *is a consultant on videodisc projects, specializing in complex Level II applications. He has served on the faculty of the University of Texas Southwestern Medical School in the Department of Medical Computer Science, where he worked initially with videodisc. He later started his own company to produce patient education materials on interactive videodisc.*

Dr. Kent received his M.D. from Baylor College of Medicine in Houston and a Master's in biomedical communications from the University of Texas Health Science Center in Dallas. He also holds a B.A. in communications and broadcast/film from Stanford University and an M.B.A. from the University of Washington.

Dr. Kent's background in film and television, coupled with his extensive knowledge of computers, makes him especially qualified to write this chapter.

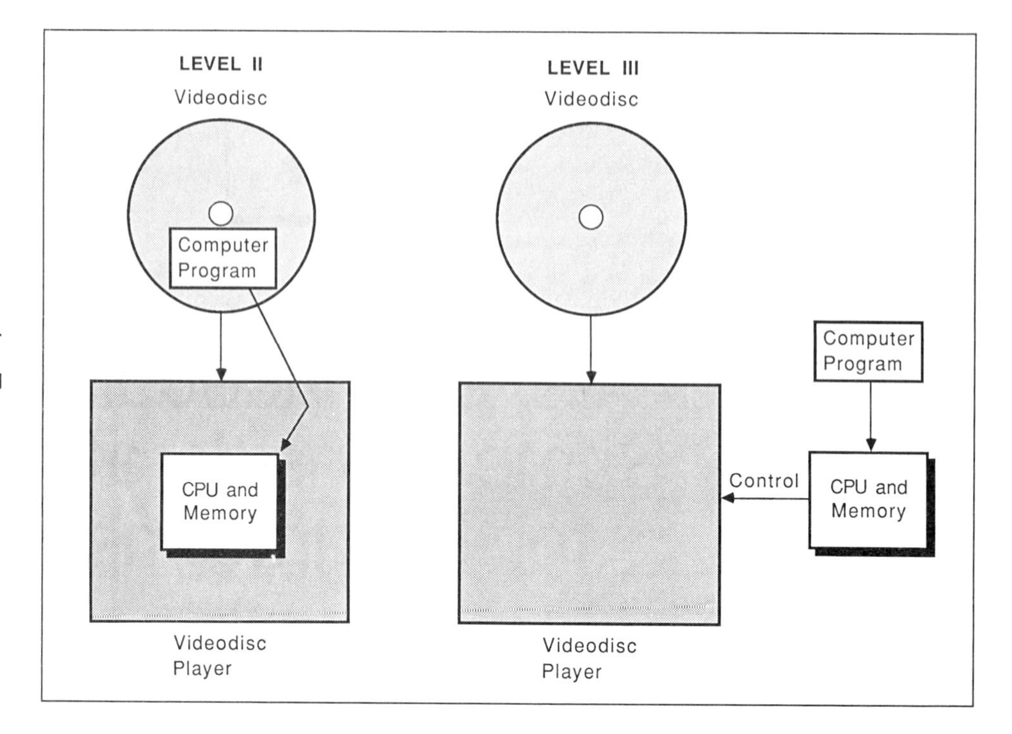

Fig. 4-1. The relationship between Level II and Level III videodiscs.

I'm going to define a *Level II videodisc* as:

> A videodisc that has a controlling program pressed onto the disc along with the video and audio signals, and that uses the videodisc player's own internal memory and processor for execution of that program.

Note that this is a hardware-based definition; the software is not defined at all, only its location for permanent storage. The definition says nothing about the nature of the program that is stored on the videodisc, only that it is stored there for retrieval and execution by the player's central processing unit (CPU).

Based on this definition, I've illustrated the relationship between Level II and Level III videodisc technology in Fig. 4-1. As you can see, the only difference between Level II and Level III is where the controlling program and the CPU are located. In Level III, both program and controlling CPU are external to the videodisc/player system; in Level II they are an integral part of it. From this perspective, it is evident that there is really no conceptual difference between Level II and Level III. They are both computer-driven videodisc systems with the potential for interactive programming. Having said that, let's take a moment to discuss the advantages and disadvantages of using Level II technology rather than Level III.

WHY USE LEVEL II? WHY NOT?

Given our working definition of Level II, why would anyone want to permanently store the controlling program on the videodisc itself and use the player's own CPU to execute it? Quite simply, because it works. Why does it work? Because it is simple.

The word "simple" here needs some clarification. Level II productions are not simple in the sense of "not complex"; they can be as complex as the designer wishes them to be. Nor are they simple in the sense that they are easy to produce; on the contrary, they are often more difficult to produce than Level III productions. They are, however, simple for the user to use, and that is their most significant benefit.

All too often, videodisc producers trivialize, or even disregard, the fact that their final product must be used in the real world. Ordinary people will have to plug in the player, put in the discs, connect the cables, turn on the power, and so on. The larger the number of people who have to handle the equipment, the more likely it is that something will go wrong. The larger the number of components in the system, the more *certain* it becomes that something will go wrong.

The beauty of a Level II system is that from the user's perspective, and compared with a Level III system, it is relatively "hardware simple." (The term *user* refers to the people who have to set up and turn on the system, as well as those who view the material presented by the system.) The computer was designed to control this particular disc player and to run this particular kind of software; the interface between the disc player and the computer was designed, built, and permanently installed by the manufacturer; and the video production and computer program are durably and inseparably packaged on the same videodisc. The user just has to turn on the power, put the disc in the player, and press PLAY. (Even the last step has been removed on some Level II systems, since they automatically go into Play mode whenever the power is turned on.)

In addition, the system is compact and standardized. If the user is making a presentation at a distant location, the player can be easily transported, or arrangements can be made to have a standard player delivered to the site. No special monitor is required, since any standard television can be used with the Level II player. The complete system can be serviced by the manufacturer's representative, without any question as to which piece of hardware is in need of service.

This end-user convenience and reliability does not come without some trade-offs, however. Level II currently has the following limitations that are not found in Level III:

1. Limited random access memory (RAM).
2. Limited input/output.
3. Limited arithmetic capabilities.
4. Slow processing.
5. A "locked-in" program.

Let's take a closer look at some of these limitations.

Limited RAM

The first Level II players, the MCA PR-7820 (later called "Discovision" and then "Pioneer PR-7820") and the Sony LDP-1000, had only 1K of addressable

RAM into which to load programs. New players have more memory (7K is now common, and one player in Europe has 45K), but they are still "memory poor" in comparison with an IBM® PC (half a megabyte) or an IBM® AT™ (2 megabytes or more). RAM limitations are disappearing as new generations of players emerge, and, in any event, this is not as great a limitation as it sounds for three reasons.

First, the operating system of the videodisc player is designed to do nothing but operate a videodisc player. Its controlling op-codes tend to be very efficient and videodisc specific, and they therefore do not take up much room in memory. For example, in Pioneer LDV-6000 language if you want to search to frame 1000 and play to frame 2000, you store the number 1000 in register 1 and the number 2000 in register 2; the data storage registers are an integral part of the program and are therefore loaded when the program is read from the videodisc. The actual program for the search-and-play program looks like this:

```
1   RECALL
    SEARCH
    AUTOSTOP
```

Once the program is assembled into its executable hex-code form, it looks like this:

```
3F 7F F7 F3
```

Only one hex code (one memory byte) is used for each op-code and each digit. The whole program uses 4 memory bytes, and 4 more bytes are used to store the two frame numbers—a total of 8 bytes. A simple BASIC program, written for an IBM PC and compiled for execution, would take up much more memory than that, because of the overhead required for sending commands to the player. Thus, the Level II languages are efficient memory users when compared with their equivalent Level III programs because the language is specifically designed to control videodisc players.

The second reason that available RAM is not a major limitation is that most programs can be divided into segments, and the various segments can be loaded and executed separately. With some players, it is even possible to do partial memory loads and thereby implement a crude virtual-memory or overlay system, holding in memory only those overlays needed for the current functions. For example, if a videodisc production is divided into multiple segments and each segment is listed and chosen from a menu, the computer program that controls segment 1 doesn't need to be in memory when segment 3 is being viewed. (Some techniques for implementing such a system will be discussed later in this chapter.)

Finally, the limited RAM of Level II players does not often limit a production's functions because, in practice, there is usually only so much you want to do on one side of a videodisc. A 7K-memory capacity is enough for 2000 to 3000 command lines—room for more functions than a 30-minute video program is likely to need. Even if you use a few hundred still-frames in the program, each frame takes up only 2 bytes of storage, so there will be ample room left for command lines. And if space does become a problem, there are programming techniques for reducing the amount of storage required for still-frames. (These techniques will also be covered later in this chapter.)

The memory restriction is a function of player design. The processors in Level II players are capable of addressing much more memory than manufacturers currently include, and, if there is a demand for it, they will undoubtedly include more. There is therefore no reason that future players should not have 64K or even 640K of memory.

Limited Input/Output

Currently, Level II videodisc players are usually designed to take input from a standard disc-player numeric keypad and to send output to a television screen (certain players can also send characters out through a serial port).

On the input side, there are no provisions for attaching a computer keyboard to the player so that it can receive and process text strings. One European-manufactured Level II player (the Philips VP835) is equipped with a light-pen interface, but such input devices as touch screens and light pens are not readily interfaceable to most Level II players. (I'll discuss the concept of external device interface again later in this chapter.) The input from the user is therefore limited to numeric input or, more commonly, to numeric selection of a single choice from a list of numbered options.

On the output side, most information is transmitted to the video screen in the form of preset video frames or video segments. Thus, any messages to the user must have been edited onto the videotape in the form of text frames or video/audio segments. The computer program itself does not store and transmit the messages; it only controls which message is displayed. There are three exceptions to this rule: First, one European player (the Philips VP835) is set up to handle teletext messages stored in the player's memory. Second, one U.S. player (the Pioneer LD-V6000) can be easily modified with an optional EPROM to permit two 16-character lines of text display. Third, if the player has an addressable serial port, then characters can be sent out to an external display device such as a printer or a terminal (this will also be covered in more detail in the section on external devices).

Limited Arithmetic Capabilities

The languages implemented on Level II players have limited ability to do arithmetic. Provisions are usually included to increment and decrement a register, and sometimes to add or subtract one register to or from another through an accumulator, but this is usually the extent of the function. One can sometimes devise crude multiply-and-divide routines using addition, subtraction, and dropping of the least-significant digit (a rough divide-by-10); however, these routines tend to be awkward and time-consuming.

The following is an example of a routine, written for the Pioneer LD-V6000, that does a rough divide-by-3 by multiplying by 0.3333. This routine is useful, for example, for determining the length (in tenths of a second) of a particular motion segment with a given number of frames, since there are 30 frames per second.

```
0         ARG     Put contents of register 0 in argument
                  register.
          DROP    Divide contents of argument register by 10
                  (multiplied by 0.1).
          GET     Put contents of argument register in
                  register 0.
R/TEMP    PUT     Also put result in temporary register.
0         ARG     Put contents of register 0 in argument
                  register.
          ADD     Add contents of argument register to
                  contents of register 0 (multiplied by 0.2).
R/TEMP    ARG     Put contents of temporary register in
                  argument register.
          ADD     Add contents of argument register to
                  contents of register 0 (multiplied by 0.3).
0         ARG     Put contents of register 0 in argument
                  register.
          DROP    Divide contents of argument register by 10
                  (multiplied by 0.03).
          ADD     Add contents of argument register to
                  register 0 (multiplied by 0.33).
0         ARG     Put contents of register 0 in argument
                  register.
          DROP    Divide contents of argument register by 10
                  (multiplied by 0.033).
          DROP    Divide by 10 again (multiplied by 0.0033).
          ADD     Add contents of argument register to
                  contents of register 0 (multiplied by 0.3333).
```

Of course, it would be much more convenient to have a divide op-code or, for that matter, just a multiply op-code. Unfortunately, the manufacturers have not provided these functions in the language. Note that the lack of these capabilities is a language (or operating-system) limitation and not a hardware restriction.

Limited arithmetic capabilities can be a drawback if Level II programs require any "number crunching" beyond simple addition and subtraction. It is hoped that manufacturers will respond to this need by including more mathematical functions in future versions of their operating systems. This modification would be relatively inexpensive, since only the player's EPROM would need modification, not its hardware.

Slow Processing

Although the processors in newer Level II players are comparable in computational speed to the processors in many microcomputers, the processing of Level II commands actually occurs relatively slowly, because of the demands on the processor of other player functions. For example, the processor in a Pioneer LD-V6000 is a Z80 micro chip—the same chip that is in a TRS-80 microcomputer. For most relevant functions, it is comparable in power to the

8088 in an IBM PC. However, in the LD-V6000 videodisc player, this processor must perform all the life-support functions of the videodisc player, such as focusing and tracking, checking field and frame numbers, and looking for commands coming in from the external port and from the remote control unit. With all these extra responsibilities, the Z80 does not have a lot of time for reading and executing the Level II commands in the player's memory.

Because it is busy with these housekeeping chores, the processor only looks for and executes Level II commands when it catches its breath between video fields. Thus, it gives the player a new command line to execute only once every sixtieth of a second. A command line consists of both the op-code and its argument. In the previous divide-by-3 routine, there are three different command lines—GET (get contents of argument register and put in register 0), PUT (put contents of register 0 into argument register), and ADD (add contents of argument register to register 0)—that are used a total of six times. ARG (used for indirect addressing) and DROP (drop least-significant digit of argument) are considered part of the argument for the commands that follow them and therefore take insignificant amounts of processing time. The total time taken to divide by three is 6/60, or 1/10, of a second.

Some commands need time to execute beyond the actual processing time; for example, a SEARCH command takes 1/60 of a second to process, plus however long it takes to physically search the target frame. However, there is no difference in execution speed between a SEARCH executed from Level II memory and a SEARCH commanded by an external computer.

The time needed to execute commands is not a problem when doing simple search, play, and user-input commands. It is also not a problem if the processing can be done while the player is displaying a still-frame or motion segment to the viewer as part of the regular program. For example, a great deal of processing can be done between the time the player first searches to a question frame and the time an INPUT command is given to permit the user to make a selection. The programmer can assume that the user will take some time to read the question and will not be aware that the keypad is inactive until the INPUT command is issued.

The problem with slow processing makes itself evident when the user is expecting something to happen, such as right after pressing a key in response to a question, and the player is spending time thinking about what it needs to do next. This can be confusing, or even irritating, if the delays are long enough. Another place where slow processing will impair the program is at the end of a motion segment; if the player must stop and think about where it should go next, a freeze frame that was not intended by the designer or the programmer will occur. These effects can be avoided by careful placement of the long processing functions before and after SEARCH commands and by judicious use of the VOFF (video off) command to avoid unanticipated visible freeze frames.

The processing speed problem will be resolved in future models of players if manufacturers use faster processors or if they include a second, inexpensive processor dedicated to handling the Level II program functions. This would also permit the expansion of control languages to include more mathematical functions, string-manipulation functions, and videodisc control functions.

A "Locked-In" Program

A concern frequently voiced with respect to Level II is that "the program is locked in plastic and cannot be changed." This refers to the fact that the controlling program is actually put on the disc during the pressing process and is thus unchangeable. In my opinion, this is not a legitimate argument against Level II, for the following reasons.

First, the program can be updated by repressing the disc. When large numbers of copies (say more than 1000) are involved, repressing can actually be cheaper than, or at least comparable in price to, replicating, distributing, and installing the new diskettes or PROMs used to control Level III videodiscs. Magnetic media and silicon chips are not free, and they both require expensive serial reproduction processes.

Second, very few applications would require updating of the controlling program without also requiring updating of the video or audio material on the disc. Consequently, it can actually be twice as expensive to update a Level III application because both the videodisc and the diskette or PROM must be replicated.

Third, once a program is released and distributed, it should rarely require updating unless the information dealt with by the program changes. In other words, programs should have been thoroughly tested and debugged before release. This principle applies no matter where the program resides; the fact that Level III programs are not "locked in plastic" does not relieve the programmer of responsibility for quality control.

Another way to avoid any problems with locking in the program is to recognize from the start that applications that do require constant updating are not good candidates for Level II. For example, it would be foolish to use Level II for catalogs, in which prices or availability are subject to frequent changes. On the other hand, an ideal application would be training programs or presentations that do not usually require this type of constant updating.

Despite these limitations, a Level II player controlled by a well-designed program is appropriate for many applications.

THE CONTROLLING PROGRAM

On the final videodisc, the controlling program is actually stored as a brief series of audio signals on the second audio track. Each series of tones constitutes one complete load of data (usually 1K) and is known as a *dump*. The player's CPU reads the audio signals and converts them into a program in much the same way that a low-cost microcomputer can receive programs from a cassette tape.

When any Level II player is first started up, it searches to frame 1, blacks out the video, squelches the audio to the speakers, and looks for an identifying tone on audio track 2. If it finds the proper tone (the frequency of the tone varies between manufacturers), it knows that it will soon encounter the controlling program, so it continues to play, listening to the second audio track. When it finds the first dump, it loads it into memory and then begins to execute the program starting at memory location 0. If the player does not find the identifying tone when it searches to frame 1, it puts itself in manual play mode and acts just like a Level I player.

Once the first dump is in the player's memory and is being executed, it can command the player to search to any particular frame on the disc and look for another dump, either to replace itself or to load into another part of memory. Since one dump can call any other dump, the player's memory is in effect expanded by as many dumps as there are on the videodisc.

The Programming Process

Knowing that the final form of the program is actually a series of audio signals on audio track 2, how does the programmer proceed from creation of the program to its final inclusion on the replicated disc? With some variations between programmers and pressing facilities, the process looks something like this:

1. The final videotape is sent to the pressing facility.
2. The pressing facility uses the tape to cut a "quick and dirty" disc called a *check disc*. (A *DRAW disc* can also be used as a check disc.) The check disc contains all the frame numbers, video, and audio that the final disc will have, but it has no program dumps.
3. The programmer uses the check disc to get the necessary frame numbers and then proceeds to write the program on a microcomputer.
4. The program is assembled on the microcomputer into a series of hex codes that are transmitted to the player's memory.
5. The program is debugged through a looping process of testing, rewriting, reassembling, transmitting, and then retesting.
6. When the program is finished, the programmer sends the assembled hex codes, or the unassembled source code, to the pressing facility, which translates the program into audio signals, adds the identifying tone, and places it on the master tape in the frame location specified by the programmer. At this point, the facility usually cuts a second "quick and dirty" disc called a *proof disc*. Because it contains the program, the proof disc has all the functionality of the final disc, but the video and audio quality may not be representative of the final product.
7. The programmer then tests the proof disc to make sure that it works exactly as it should, at which point the disc is released for replication.

Several of the steps in this process are self-explanatory, but a few bear some further discussion.

Writing the Program

Each brand of player has its own set of op-codes and its own way of reading information from the audio track. Different models within the same product line may not be able to understand each other's op-codes. Therefore, it is imperative that the programmer know right from the start on which player or players the program will be expected to run. The programmer then writes the program in a player-specific language that looks a little like a cross between assembly language and video-editing language.

To complicate matters, the programmer must write not only for a specific player or players but also for a particular assembler. Pioneer, Sony, and Philips all sell software packages for microcomputers that can be used to write and assemble programs for their respective players (Philips's program, like its Level II player, is available only in Europe), but the assemblers in these packages are not interchangeable. There are more-generic packages available from other sources, though, and some programmers have opted to write their own assemblers to get around the compatibility problem.

The following program is a segment of one written for the Pioneer LD-V6000 in the syntax of my own assembler (the commands are identical to the ones that Pioneer uses in its assembler). This example is just to give you a feel for a player-specific language, so I'm not going to explain it in detail.

```
QSUB:    R/QOFIN   PUT      R/QOFIN is a pointer to register stack.
                            Local variable.
         1         ARG      Protect return address for nesting of GET
                            subroutines.
         R/QPR     PUT
USERIN:  R/QOFIN   ARG      Isolate search register for question
                            frame, starting with pointer to register stack.
                   RECALL
                   SEARCH   R/QOFIN + 1 should now be active.
                   VON
                   ARG      Use active register or R/QOFIN + 1,
                            which has number of answers in frame,
                            and search register of segment to play
                            for Action Indicator = 1.
                   ARG
                   GET
         0         ARG
                   DROP
         0         SUB      Leave number of answers in register 0.
         1         INPUT    Allow user input of 0 (previous menu) or
                            other.
```

Compare this program with a segment of one written for the Sony LDP-1000 in the syntax of an assembler available from Videotools, Inc.:

```
EXERC5  R= REG0 4 *SET REGISTER FOR MENU ON MENU KEY
                  CH ONE ON
                  CH TWO OFF
                  R= REG2 0          *SET FOR REVOLVING RESPONSES
                  R= REG3 1          *SET TO COUNT QUESTIONS
                  S- M-5-A           *PLAY SEG M-5-A
Q:S-5-01          S- S-5-01          *ENDS ON FRAME S-5-01
                  INPUT 2            *ALLOW TWO USER INPUTS
                  R:S-5-02           *1 IS RIGHT, SHOW CORRECT FRAME
                  WRONG-4            *2 IS WRONG, SHOW REMEDIATION
R:S-5-02          S- S-5-02 STOP 3 *3 SECONDS 1ST ONE; 2 THEREAFTER
                  INC REG3           *NEXT QUESTION #
Q:S-5-03          S- S-5-03
```

```
                INPUT 2
                R:S-5-04
                WRONG-4
R:S-5-04        S- S-5-04 STOP 2
                INC REG3
```

Notice that both languages, although entirely incompatible, resemble computer assembly language and are not self-explanatory. Good programmers are conscientious about providing generous comments in the source code to compensate for this.

Also notice that in both examples I have made extensive use of symbols in lieu of both frame numbers and register numbers. This makes debugging and changing the program far easier. I usually include a table at the beginning of the program that sets all the symbols to their respective values.

Transmitting the Program to the Player

Once the program's source code has been run through the assembler, the resulting hex codes must be transmitted to the player's memory so that the program's ability to control the check disc can be tested. There are two practical ways to transmit the program.

First, the program can be sent to the player through the player's communication port. This requires a standard communication link between the programmer's computer and the disc player. The link must be two-way, since the computer must know when it is clear to transmit data. Most commercially available assemblers include this type of transmission utility and can handle all the communication protocols for the programmer. Once the utility has loaded all the locations in memory, all the programmer has to do is issue a RUN command, either from the keypad or again through the communication port. The program can then be tested.

One disadvantage of transmitting the program to the player through the communication port is that this is not the way the player will receive the program when it is finally pressed onto the disc. Some players reset the status of certain functions when the RUN command is issued externally, but they do not reset them when a load occurs from a disc. As a result, some problems may not be apparent during debugging. For example, one program dump might turn off the video before searching for the next dump to load (this is, in fact, a very common practice in multiple-dump programs). Even if the programmer forgets to turn the video back on in the second dump, the video will still be reset to the ON state by a RUN command, and thus the program appears to function correctly during the debugging process. Unfortunately, when the proof disc returns from the pressing facility and is tested, it will start the second program dump with the video turned off, and it will continue to operate with the video off until it is turned back on by a VON command. At least one software package (Pioneer's) gets around this problem by storing the status of the resettable functions before loading a new dump and then returning them to that status after the RUN command is issued.

The second way to transmit a program is to translate the dumps into the same type of audio signals that will eventually be pressed onto the disc, and then feed them into the player's circuitry at the same place that the player would

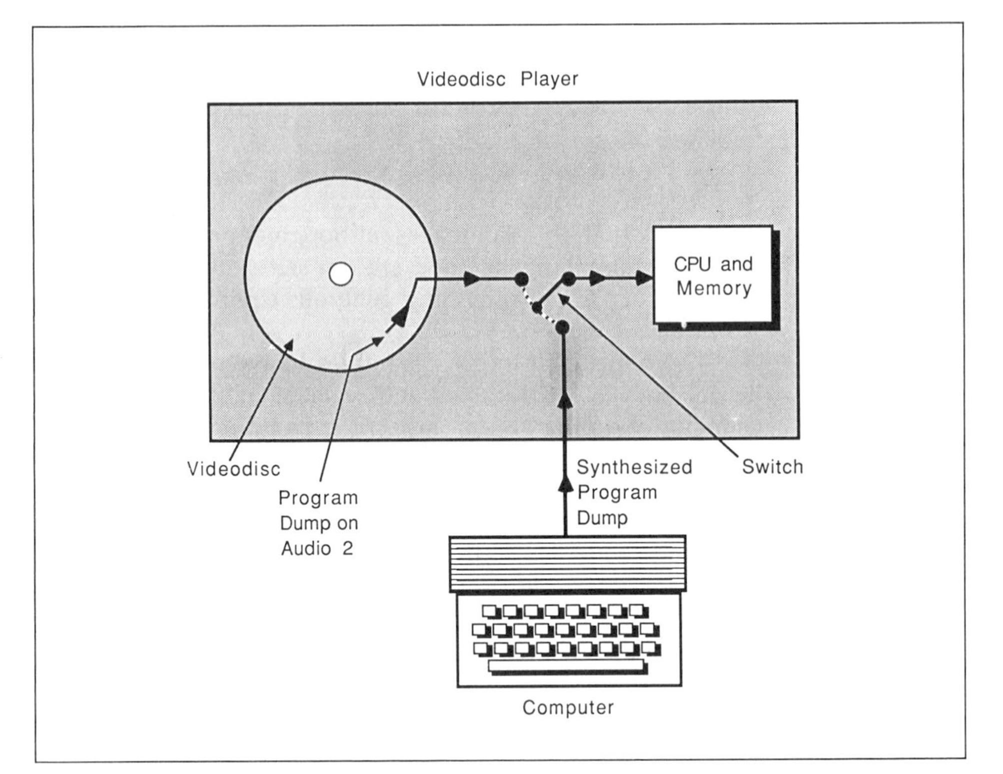

Fig. 4-2. Transmitting a program to the player by feeding audio signals into the player's circuitry.

receive the signals if they were coming from the audio track of the disc. This concept is illustrated in Fig. 4-2. In essence, you are fooling the player into thinking that it is getting its program dump from the audio track of the disc, rather than from an outside source, so that it behaves during testing exactly as it would when playing the actual disc. Of course, this alleviates the problems caused by resetting the functions, since the external RUN command is never issued.

Unfortunately, this technique requires modifying the videodisc player in order to feed in the signal. However, the modification is usually a minor one, requiring only the cutting of a trace on a circuit board and the soldering of two bypass wires. A switch can even be installed to permit changing the player between accepting audio-signal dumps from the disc and from the programmer's computer.

Sending the Program to the Pressing Facility

Once the program's dumps have been thoroughly tested and debugged, they must be sent to the pressing facility for inclusion on the audio track of the disc. This can be done in one of three ways.

First, a printout of the source code can be sent to the pressing facility, where it is manually keyed into the facility's assembler/translator. The facility usually charges extra for this type of program transfer, if they are willing to do it at all. This method is generally not recommended. It is prone to transcription errors and is time-consuming, thus tending to increase turnaround time. The only situation in which this kind of transfer is appropriate is if the programmer has no way to translate the program into the pressing facility's assembler format or hex-code format.

The second way to transmit a program is to send a diskette containing the source-code files to the pressing facility, or to send the files via a modem. For this to work, the programmer's assembler must use exactly the same syntax as the pressing facility's. This technique has the advantage of being completely electronic and therefore not subject to human errors.

The third technique is to send a diskette containing the hex-code files to the pressing facility, thereby bypassing the pressing facility's assembler. For this to work, the hex file must be in the exact format specified by the pressing facility; but, since it is usually simple to write a short utility program that will convert one hex format into another, it doesn't really matter what hex-code format the programmer's assembler creates.

The following two examples are portions of a program written for the Pioneer LD-V6000 player and assembled first in Pioneer and then in 3M hex-code format. I have omitted the same number of bytes from the middle of each example to show the header and trailer in each format.

```
PIONEER FORMAT:    ;*$PROTECT 20
   ; album # 12-123-1234          Date: 06-25-1986
   ; Prog 9, Side B               Dump # 3 Of 4
   ; Customer Name                ; 19680 Frame # of dump
   ; Rick Kent, M.D.                   Source: POPT-9B3.VD
   ; 206-745-5400                      Object: POPT-9B3.HEX
   .2F .AF .5F .CF .8F .2F .3F .CF .0F .3F
   .09 .0F .0A .08 .0F .0F .09 .0F .3F .0A
   .7F .F7 .1B .0F .3F .0A .0A .08 .9F .09
   .0F .1F .08 .5F .09 .0A .0A .08 .3F .0A
   .1D .3F .03 .0F .F8 .2F .4F .CF .1C .1E
   .04 .6F .AF .CF .6F .AF .CF .8F .0F .2F
   .0B .0F .9F .AF .CF .1E .08 .0A .02 .3F
   *************removed*****************
   .18 .13 .06 .D9 .18 .13 .07 .41 .53 .C2
   .12 .F5 .07 .3D .07 .3D .07 .3D .06 .D9
   .06 .D9 .06 .D9 .07 .41 .52 .C6 .12 .F5
   .07 .3D .07 .3D .07 .3D .07 .3D .07 .3D
   .07 .3D .06 .DD .52 .C3 .06 .D6 .52 .C0
   .1B .76 .5E .BF .07 .3A .5E .C5 .2B .90
   .00 .00 .00 .1E .00 .00 .00 .00 .00 .00
   .00 .00 .00 .00
   ;*$CHECKSUM AD

3M FORMAT:    :01270F000300
   :0103EC0003000
   :0A0000002FAF5FCF8F2F3FCF0F3F00
   :0A000A00090F0A080F0F090F3F0A00
   :0A0014007FF71B0F3F0A0A089F0900
   :0A001E000F1F085F090A0A083F0A00
   :0A0028001D3F030FF82F4FCF1C1E00
```

```
:0A003200046FAFCF6FAFCF8F0F2F00
:0A003C000B0F9FAFCF1E080A023F00
*********removed**************
:0A03A200181306D918130741 53C200
:0A03AC0012F5073D073D073D06D900
:0A03B60006D906D9074152C612F500
:0A03C000073D073D073D073D073D00
:0A03CA00073D06DD52C306D652C000
:0A03D4001B765EBF073A5EC52B9000
:0A03DE000000001E00000000000000
:0403E8000000000000
:004CE00100
```

In the Pioneer example, supplementary information is included in the comment lines (which begin with a semicolon) at the beginning of the file. In the 3M example, some of this information is encoded as actual hex lines. For example, *;*$PROTECT 20* in the Pioneer file specifies that what follows is only a partial load of data that will not fill the entire memory. In this case, the number 20 will be subtracted from 1024 to determine the number of bytes (1004) to load. In the 3M version, this information is encoded into the second line as *3EC,* which is 1004 in hex format. The frame number at which the dump should be placed, 19680, is specified in the third comment line of the Pioneer file; whereas in the 3M file, it appears in the last line as *4CE0,* which is 19680 in hex.

The actual data on each line of the two files is identical, except that 3M's format also requires that the first two characters after the beginning colon specify the number of data bytes on the line (0A in all but the last line in this example), and that the next four characters specify the starting address of the data in the line.

Obviously, the programmer has to be very careful to obtain exact data format specifications from the pressing facility. These specifications change from time to time, and no programmer wants, when working under a deadline, to have all files rejected because the format is not correct. Just as obvious, when using this technique to send a program to the pressing facility, the programmer must ensure that the hex-code file works perfectly in conjunction with the check disc and the player.

Even with these notes of caution, I feel that sending hex-code files is the preferred method for sending programs to the pressing facility, since it eliminates the problems that can occur should the programmer's assembler be even slightly incompatible with the pressing facility's assembler.

Programming Considerations

Although it is beyond the scope of this chapter to provide a complete programmer's manual for Level II players, I would like to cover a few general techniques. These include:

1. The use of subroutines.
2. Looking beyond the programming manual.

3. Passing information from dump to dump.

The Use of Subroutines

As anyone who has written even simple programs knows, *subroutines* are sections of code that can be called on to perform predetermined functions. They can be incorporated into Level II videodisc programs as easily as any other program.

For example, I use two subroutines in almost all of my programs. One subroutine handles all the motion segments. By calling this subroutine and passing it the numbers of the registers containing the first and last frame numbers of the motion segment, my programs permit the user to interrupt the motion segment in order to do such things as back up 10 seconds, skip ahead 10 seconds, pause and restart, return to the beginning of the segment, skip to the end of the segment, return to the last question, or go to the main menu.

The second subroutine I use handles all questions and options (that is, all still-frames with user inputs). The data passed to this subroutine includes the search frame of the question itself, the number of possible user responses, and the action to take on any valid user response. The data specifying these actions is also stored in a stack of registers. Possible actions include branching to a memory location, adding 1 to a score register and then branching, playing a motion segment or series of motion segments and still-frames along with various combinations of adding or not adding 1 to the score register, returning to the question after the segments, or exiting the subroutine.

Why would a Level II videodisc programmer use subroutines like these? There are three reasons:

1. Subroutines reduce the need for debugging.
2. They make the program far easier to modify.
3. The use of subroutines can conserve precious memory.

Reduced need for debugging: Once a subroutine is written and fully debugged, it should not have to be tested further and debugged unless it is changed. Thus, as long as the programmer is consistent in the way data is passed to the subroutine and the way the subroutine is called, the subroutine will work correctly every time. The custom part of each program then becomes a series of calls to the proper subroutines and very little custom code needs to be written. The new code is in turn easier to read and follow, and the fewer the lines of new code in a program, the smaller the debugging task.

Ease of modification: Subroutine-based programs tend to be data driven. In my subroutines, for example, the actual instructions for how to handle any specific motion segment or question are stored as numbers in registers. These registers can be set up at the beginning of the program, so that changing the program consists of merely finding the proper register and changing the five-digit number stored there. (Because a register is 2 bytes long, it can store a number from 00000 to 65535.)

A word of caution is in order here, lest this process sound easier than it really is. To use the registers efficiently, it is possible to store more than one

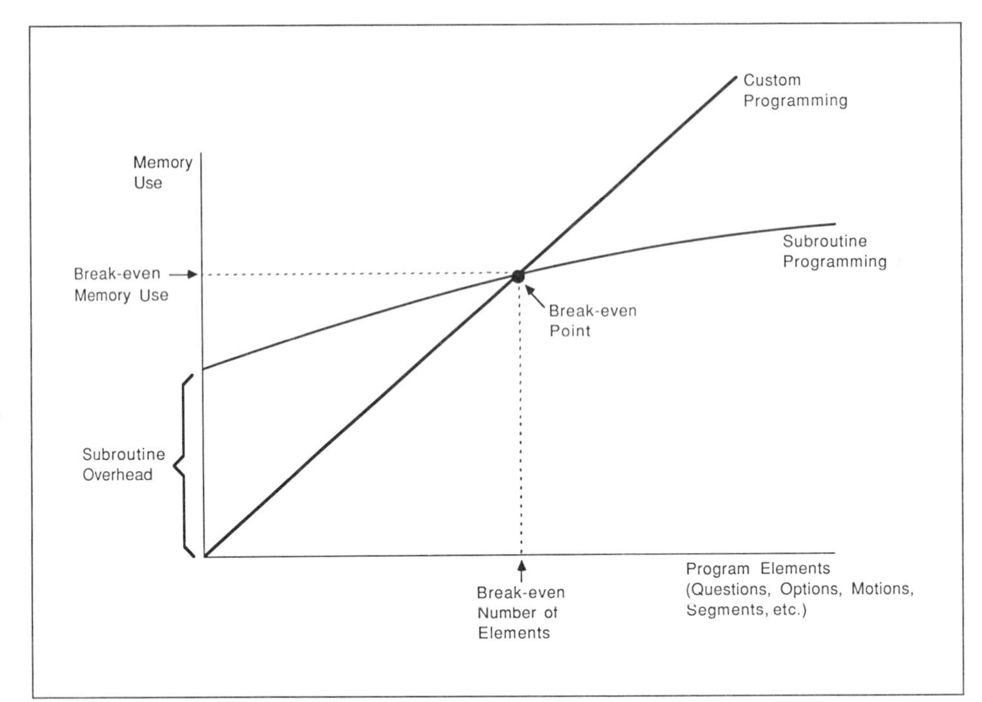

Fig. 4-3. A comparison between the memory used by subroutines and that used by other program elements.

piece of information in each of them. Memory locations are never more than four digits in length on 7K machines (7167 is the highest address), so one digit remains in a register used to store a memory location. This digit can be used to store a one-digit action-indicator code, which then tells the subroutine what actions to take in addition to branching to the specified memory location. This type of data compression, though efficient, can be difficult to set up and interpret, and therefore difficult to change without introducing errors. A branch to memory location 2535 with an action indicator of 4, for example, would appear as 25354 in the register. Because it is not readily apparent what this number represents, it could be mistaken for a frame number or some other combination of data elements.

Conservation of memory: The final reason for using subroutines is *potential* conservation of memory. I emphasize *potential* because a subroutine-based program does not always take up less total memory than a program that performs the same functions without subroutines.

Subroutines are overhead. Before the programmer has even started the custom part of the program (the calls to the subroutine and the data registers), the subroutines have consumed a major portion of memory. However, the memory used by a subroutine call combined with the memory used by the data registers is far less than the memory used by a custom program segment; so, sooner or later, the memory savings makes up for the memory overhead of the subroutines. After the break-even point is reached, the memory-saving benefits of using subroutines continue to accumulate. This effect is illustrated in Fig. 4-3, which graphs the memory used by subroutines against that used by program elements such as questions, options, still frames, and motion segments.

The break-even point is highly dependent on the nature of the subroutines, the efficiency of data compression in the registers, and the number of times the different subroutines are called. My experience is that the break-even point usually occurs around 1K of memory. This means that on a 1K machine it is usually a toss-up as to which technique uses less memory; one would almost have to program each dump both ways to truly ascertain the more efficient technique. However, focusing solely on memory ignores the benefits of less debugging and ease of modification, which from my point of view tilt the balance in favor of using subroutines.

The only disadvantage of subroutines is that they generally require more processing time than a piece of custom code to perform the same functions. This is to be expected, since a subroutine must be able to do more than the custom code, and the process of "deciding" what must be done takes time.

Looking Beyond the Programming Manual

The manuals that come with the various players list the Level II op-codes, describe each op-code's function, and sometimes give short programming examples using the op-codes. It is important for the programmer to keep in mind, though, that the functions are not limited to the uses envisioned by the player's designers. The op-codes can often be used in creative ways never intended by the player's designers.

For example, Pioneer introduced the LD-V3000 8-inch disc player with only a subset of the LD-V6000's op-codes. One of the codes left out of the new player was RND, the random number generator. On the surface, this would seem to eliminate the possibility of branching on a random basis. However, although the player has no internal random number generator, it does have an external one: the user. One of the op-codes included in the subset is CLOCK, which starts a clock that counts in tenths of a second. When another CLOCK command is issued, the clock's current value is placed in register 1 and the clock is restarted from 0. Since each user generally takes a different amount of time to answer a question, starting the clock upon reaching a question and stopping it after the user presses a key produces a number that, if only the least-significant digit (or possibly the two least-significant digits) is used, is effectively random. This is usually enough to meet the needs of most programs, which generally offer choices between fewer than ten branches.

Other examples of creative programming abound in Level II applications, as might be expected when manufacturers give programmers a limited set of tools with which to work. The important thing to remember is that just because a function is not in the manual does not mean that it cannot be done.

Passing Information between Dumps

Earlier, I mentioned that some players permit the programmer to implement a crude virtual memory, or overlay system, by partial loading of memory. The key to this capability is the ability to pass information between dumps. If one dump could not pass information to another, there would be no way for an incoming dump to know which dump called it, where it should start executing, what the status of various scores and registers was, or where it should go when it finished executing. By reserving a few registers and only loading the memory below them, the programmer can preserve such things as the memory location at which to start execution, or the frame and memory location to which the programmer should return after executing the new dump and reloading the old dump (assuming that the first dump needed the temporary services of the second dump). These registers become, in effect, global registers, whose contents can be examined or modified by any dump, but which are not changed simply because a new dump has replaced an old one.

This overlay technique is much more important for 1K machines than it is for 7K machines, since only one 1K dump can reside in memory at a time. On a 7K machine, the full memory can be considered to be seven dumps long, and each 1K dump to be a partial load of memory. In this case, it is wise to reserve the entire last dump for register data and the earlier dumps for program commands and subroutines.

LEVEL II DISC GEOGRAPHY

The principles involved in structuring audio/visual material for Level II and Level III videodiscs are the same: Searches should be as short as possible, global menus should be placed in the middle of the disc (or in the middle of the section from which they can be called), and so on. However, a few rules of thumb can be applied specifically to Level II productions, most of which relate to the conservation of memory.

First, there are some optimum places on the disc for text still-frames, which cannot be generated by an outside source in Level II as they can be in Level III. If a motion segment is immediately followed by a question or other type of text still-frame, the still-frame should be the last frame of the motion segment with the text added over that during the postproduction process. This preserves memory in two ways: The register that contains the location of the still-frame can be the same register that contains the location of the last frame of the motion segment and, depending on the type of question or option, it may be possible to eliminate the op-code for searching to that frame altogether. In any case, this type of frame layout makes for a smoother look to the production, since there are no hard searches after motion segments.

Second, savings in both programming time and memory space can be achieved by consistently placing certain frames in the same relative position to certain other frames. For example, an application's menu might include sections of the production that are located on the other side of the disc. When the user selects one of the sections from the menu, a frame that reads "For this selection, turn the disc over" might be displayed. Since menus should usually be placed in several locations on a disc (the beginning, middle, and end), the "For this selection . . ." frame should immediately follow the menu frame, wherever the menu frame is located. That way, even though the menu frame occupies a register, the "For this selection . . ." frame can be accessed with just a step-forward command (or three step-forwards, if three copies of every still-frame are included on the disc as a precautionary measure, as is the custom). The code for doing the step-forward will be exactly the same in each dump because the relative location of the two frames is the same.

Third, it is possible in certain contexts to arrange still-frames so that they actually extend memory. For example, an application might include on the disc a dictionary of the technical terms it uses. When the user asks to see the dictionary, a frame with a list of terms is shown, the user selects a term by pressing a key, and a brief definition is displayed. Each definition is one text frame, and, by sequencing the definition frames after the dictionary frames in the proper order, it is possible to avoid directly addressing them. Instead, the value of the key pressed is added to the value of the dictionary frame number (or is added to it three times, if there are three copies of each frame), and then the player searches to the resulting value to find the proper definition frame. The alternative is to store the frame numbers of each definition within the program, thereby taking up valuable space. Other types of still-frame sequences can be handled in this way, which in effect creates a branch of memory at the location of the anchor frame (in the example, the dictionary frame).

Finally, the question arises as to where to place the digital program dumps on Level II videodiscs. The first dump must always occur at frame number 1, but after that it is up to the programmer to locate them. Let's first look at the few places where they cannot go.

Since the dumps are actual audio signals on audio track 2, they cannot be placed anywhere on the second channel where there is other audio. Next, extreme caution should be exercised when placing dumps with frames that have actual visual images on them (as opposed to video black) because, even though the players black out the video automatically once they begin to load a dump, they must first search to the dump location, and the screen will flash as the player lands on the dump frame. On players that permit the programmer to blank out the video, this can be avoided by issuing the VOFF command immediately prior to searching for the dump frame. However, because these players (with certain exceptions) tend to search back to the original dump frame before beginning execution again, the video should not be turned back on until after the search to the first frame in the new dump begins; otherwise, there will be another flash when the VON command is issued.

The best location for program dumps is usually determined either by the calling dump or by the dump that is being loaded. One search can be avoided by placing the dump next to the location from which the call for it is likely to occur, or by placing it at the first frame to which the player is likely to search. Suppose, for example, that an application uses a global Help function; that is, Help can be called from anywhere in the application. The first thing the user sees after pressing the Help key is the Help menu. If the Help function is completely controlled by one program dump, then that dump should be placed next to the Help menu on the disc. That way, when Help is called from another dump, the video is turned off, the player searches to the Help menu, the adjacent dump is loaded, the player automatically searches back to the Help menu, and the video is turned back on. This eliminates having to search to the dump that controls the Help function.

Another good location for program dumps is next to menu frames. Since different selections from a menu are usually controlled by different dumps, locating one of the dumps next to the menu saves the search to that dump when the selection is made. Any other dumps that will be called as a result of a selection from that menu should be placed as close to the first dump as possible to minimize the search time required.

When placing dumps one after the other, space must be allocated not only for the program dump itself but also for its leader tone. The pressing facility will usually specify the minimum number of frames to allow for program dumps.

BEYOND THE BASICS

Two additional topics warrant some mention in this discussion of programming for Level II videodiscs: connecting external devices and creating a disc that will function at both Level I and Level II.

Connecting External Devices

Because most videodisc players have external communication ports, it is possible to connect certain devices to the player so that data will travel to a limited extent in either direction. Candidates for connection include a plotter or printer, a computer, or a modem. It is even possible to connect a player to one or more other players, in effect creating a master controller player with a number of slaves, or to design custom intelligent devices ("black boxes") that can perform any number of functions, from controlling servos to interfacing with a light pen or touch screen. Although in a Level II system the communication back to the player is limited to inputs from the user's keypad (0 through 19), for many applications this may be enough.

The major limitation to this type of enhancement is the player's operating system, since most players are not set up for sophisticated data communication and the transmission rate is fairly slow. The maximum speed is 60 characters per second because, as discussed previously, the processor can process only one command line every sixtieth of a second. Actually, this speed is a theoretical maximum that might possibly be achieved with a series of TM (transmit) commands like these:

```
BEGIN:    MEMLOC    STP  Set transmit pointer to memory location of
                         message.
          TM             Transmit at transmit pointer.
          ITM            Increment the transmit pointer and transmit.
          ITM
          ITM
          etc.
```

The shortest loop that could be written for the LD-V6000 player would look something like this:

```
BEGIN:    MEMLOC    STP  Set transmit pointer to memory location of
                         message.
LOOP:     ITM            Increment the transmit pointer and
                         transmit.
          LOOP BRANCH    Back to beginning.
```

This totally impractical infinite loop takes 1/30 of a second to transmit a character: 1/60 for the increment and transmit, and 1/60 for the looping branch. Including a countdown of the number of characters (locations in memory) to send, so that the program will exit the loop when the entire message has been sent, produces a routine something like this:

```
BEGIN:   MEMLOC    STP      Set transmit pointer to memory location
                            of message.
LOOP:              ITM      Increment the transmit pointer and
                            transmit.
         R/COUNT   DECREG   R/COUNT started out with the number of
                            characters to send. DECREG will decrement
                            the register and skip the next line when
                            R/COUNT = 0.
         LOOP      BRANCH
         EXIT      BRANCH
```

This more realistic program sends a character every 1/20 of a second because three commands must be executed for each character sent. This rate is the equivalent of about 200 baud.

Why go to all the trouble to design and interface a piece of external intelligence to a Level II player? The reason is that nearly all the benefits of the Level II system can be maintained while expanding the power of the system. As long as the intelligence in the device is designed to be general in nature, rather than program-specific (as it is, for example, in a printer), then the reliability and simplicity of the system can be maintained. Each disc will carry its own program from which it cannot be separated.

Level I/Level II Discs

From a technical perspective (though perhaps not from a stylistic one), any disc can be used as a Level III disc. It is only necessary to put the disc in a player, connect the player to a computer, and take control from that outside source, ignoring or bypassing any materials that are specifically Level I or Level II. On the other hand, the only discs that can be used as Level I or Level II discs are those that have been prepared specifically for these purposes.

It is possible when planning a disc and laying out its materials to produce a disc that will work equally well in either Level I or Level II systems. Essentially, this is done by careful mapping of chapter frames and picture stops (Level I features) and of program dumps and still-frames (Level II features).

An example of such a layout is the program illustrated in Fig. 4-4, which lends itself to chapter segmentation and option selection from a menu. The black frame at frame 1 permits Level II players to find the first dump without causing a flash, while the picture stop on the Level I menu at frame 2 causes Level I players to stop without playing over the Level II dump tone (at most, the tone will be heard for 1/30 of a second as the player plays over frame 1). Thereafter, the Level II still-frames are sandwiched between the Level I picture stop menu at the end of each chapter and the chapter stop at the beginning of

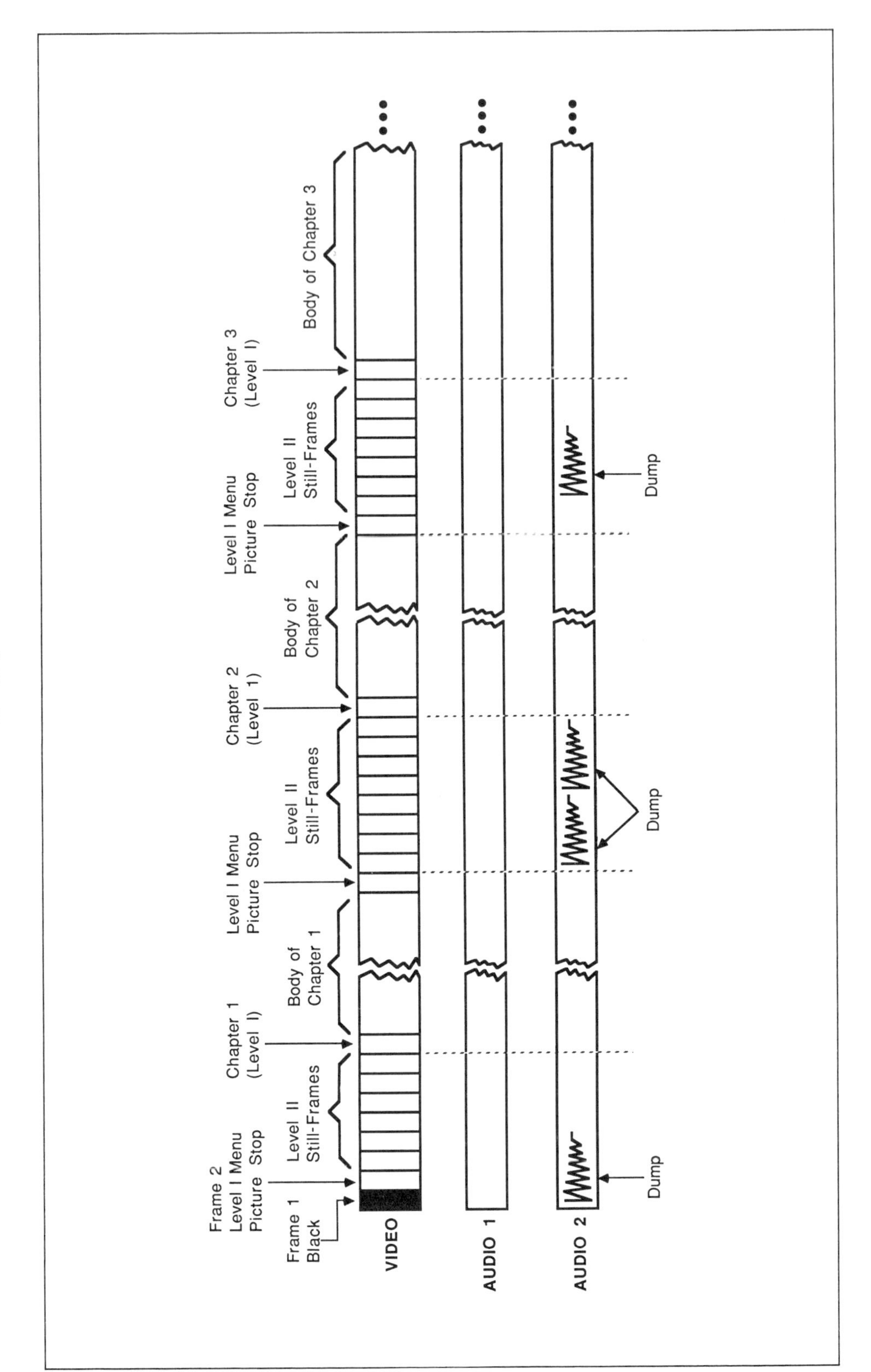

Fig. 4-4. The layout of a program for a videodisc that can be used on Level I and Level II players.

the next chapter. Any additional program dumps are located on audio track 2 after the Level II still-frames.

A Level II player will not recognize the picture stops or chapter frames of the Level I part of the disc; so, even if it plays over one, which it should not, the Level II presentation will not be affected. Level I users will not see the Level II still-frames or hear the dump tones as long as they select their options from the main menus, because the Level I player will always stop before it reaches any Level II frames and will search over them when a chapter number is selected.

Of course, this type of layout will work only with certain types of Level I or Level II programs. However, it is often worth a little extra effort and creativity to create a videodisc that can be used on as many players as possible.

A PARTING COMMENT

Level II offers a set of tools that are appropriate to some jobs and not others. Misconceptions about the relative merits of Level II stem from the use of the word *level*, which commonly implies value. After all, it is rare that *high level* is not worth more than *low level*, and in our "if one is good, two must be better" society, people tend to assume that Level III must have a higher rating than Level II, which must in turn have a higher rating than Level I.

Although the connotation of higher rating for Level III productions was strictly semantic in origin, it has often been assumed that in the world of interactive videodiscs a higher rating must imply a higher degree of interactivity. It is, after all, the practical potential for interactivity that sets the videodisc apart from film and videotape. Consequently, there is a common misconception that Level III productions are inherently more interactive than Level II and, by inference, more complex and therefore better.

There is, however, no inference of interactivity or complexity in the Level I/Level II/Level III classification system, and it is easy to imagine a Level I production that is much more complex and much more interactive than a very simple Level III production. The potential for complexity and interactivity resides in the hardware delivery system, but the extent to which a particular production makes use of that potential does not. So, as a final comment, I would like to stress that, when considering production of a videodisc, it is important to concentrate on what needs to be done and to let the application determine the choice and use of the hardware system, rather than the other way around.

5

A Survey of Level III Videodisc Delivery Systems

Richard Haukom and Eric Malone

Richard Haukom *is President of San Francisco–based Haukom Associates, an interactive consulting and design firm, whose clients include Safeway Stores, Science Research Associates, and One Pass Film and Video. Mr. Haukom holds degrees in philosophy and instructional design, and is currently working toward his Ph.D. at Stanford University. He is President of the International Interactive Communications Society.*

Eric Malone *runs his own film and videotape production firm in San Francisco: Malone & Tessman Productions. His most recent work includes a training videodisc for Hewlett-Packard on local area networks. Currently, he is designing a sophisticated Level III interactive videodisc game. Since receiving a B.A. in film and video production from Indiana University, Mr. Malone has acquired more than twelve years of experience in broadcasting and video production. In addition, he is a published author of fiction, poetry, articles, and screenplays. Mr. Malone is a member of the Membership and Special Projects Committees of the International Interactive Communication Society.*

The authors' experience reflects the special blend of talents needed in this field, just as their chapter reflects the special blend of hardware and software needed for Level III applications.

A common misconception when discussing the various levels distinguishing interactive video systems is that Levels I through III refer to the degree of interactivity available in the system. Instead, the terms *Level I, Level II,* and *Level III* are intended to indicate the locus of control over the video program. Thus, in Level I, the control lies with the user and is executed by means of the remote control device. In Level II, the sequence of events is determined by the internal microprocessor and its downloaded code, leaving the user to respond to the content of the program and not to its mechanics. Level III relocates the control functions to an external computer system, thereby providing designers of interactive discs with more options for input, data storage, and complexity of program function. The terms *Level IV* and *Level V* have been coined to suggest fundamentally different capabilities than Level III, but, since the locus of control remains with the external microcomputer, it is our opinion that these levels should be described merely as enhanced Level III systems.

Why link a laser videodisc player to a computer? Because with a sophisticated microcomputer running an application, far more flexible communication can take place between the user and the system. The same video material can be used for different purposes employing different control programs. Information regarding user input can be stored, analyzed, and transmitted. Also, the addition of the computer to the videodisc allows instructors and designers to incorporate dynamic screen elements via computer-generated text and graphic overlays, so items that change frequently, such as prices or performance measures, can be quickly and inexpensively updated and displayed.

Viewers will often find that the Level III environment is more hospitable, since they can respond to Level III presentations by using a variety of input devices such as light pens, touch screens, or specially designed sensors (for example, the "Resusci-Annie" mannequin used for CPR training). Further, Level III systems can easily accommodate many other peripherals, such as printers or modems used to relay information on sales transactions or equipment maintenance.

Level III systems also offer designers the option of using both a video monitor and a computer screen to supply the user with separate image and text displays—or both displays can be combined on one screen using *graphic overlay* techniques that place computer text on top of the picture from the videodisc.

This chapter examines Level III system components, such as videodisc players, computers, interfaces, and authoring software. It also discusses graphic overlays and compressed audio and data, and describes a few of the many Level III applications.

LASER VIDEODISC PLAYERS

Setting up a Level III interactive videodisc (IVD) system requires careful consideration of which laser videodisc player to use. Most machines are capable of being interfaced to computers; some can even work in both Level II and Level III modes. The primary suppliers of videodisc players in the interactive field are Pioneer, Sony, Hitachi, and Philips. Others, such as Magnavox, Sylvania, and TEAC, are remarketing consumer machines from Pioneer. Of all the different players, Pioneer and Sony are the most widely used in the United States.

Earlier models of both Sony and Pioneer players lacked the weight and size advantage of the current generation. Because they used helium-neon, or "HeNe," gas lasers, which required larger power supplies than the current solid-state (or diode) lasers, these early models were somewhat heavy and bulky. All current players are relatively lightweight and compact, and have standard functions, including forward and reverse play, still-frame, fast forward and fast reverse, scan, slow motion, picture stops, chapter stops, and frame-search capacity. In addition, most players have standard serial or other communications ports so that they can easily be linked to a computer.

One consideration in model selection is the speed with which the player can access any frame on the disc. Early players took as long as 19 seconds to search from the first to the last frame. Today, some players can perform this task in as little time as 1.5 seconds. Most of today's players, however, have an average end-to-end search time of about 3 seconds.

Sony Laserdisc Players

Sony offers a line of laserdisc players built to work with Level II and Level III applications. The Sony LDP-2000 is the current series, offering an RS-232 (serial) port and an optional IEEE-488 (parallel) port. Its optional on-board microprocessor can access up to 7K bytes of data for Level II programs. Currently, this machine has the fastest access time: 1.5 seconds or less. The LDP-2000 also has options for still-frame audio and computer-data capture from digital information stored on the videodiscs. The LDP-2000 is designed for modular expansion and, depending on the type of peripherals added, its cost can range from $1900 to $2500.

Sony's previous model, the LDP-1000A, has been discontinued, but it is still available through distributors for about $650. The 1000 series employed a helium-neon laser and contained 5K of random access memory (RAM) to serve its Z80 microprocessor for Level II applications. With its RS-232 port and current pricing, the LDP-1000A is an inexpensive, industrial-grade machine, especially suited to long-running stationary applications where its size, weight, and top-loading design do not prove significant drawbacks.

Pioneer Laserdisc Players

The Pioneer LD-V6000 and its derivatives (LD-V6010 and LD-V6200) are also capable of running both Level II and Level III applications. The LD-V6000 and LD-V6010 players provide an RS-232 interface for serial communication, and the LD-V6020 comes with an IEEE-488 parallel port. They each store 7K of data in RAM for use by a built-in Z80 processor for Level II programs. The 6000 series is front-loading and has a 3-second maximum search time. The various models are priced for the industrial market at costs of $1500 to $2000.

The discontinued Pioneer LD-700 is a consumer-level laserdisc player capable of Level III operation with a computer interface through an eight-pin DIN jack. One drawback to the LD-700 and other consumer models is their inability to send *frame-status information* (the number of the frame that the disc player was on when the user acted) back to the computer. Prices on this model have dropped to around $400 at many consumer outlets.

Pioneer's CLD-900, priced as low as $700 at some retailers, is capable of playing either digital audio compact discs (CD) or videodiscs containing digital or standard analog audio. In 1986, Pioneer issued the CLD-909, also designed to attract consumers interested in the burgeoning CD market. It is generally felt that the unexpected surge of demand for compact disc players bodes well for future consumer acceptance of videodiscs. The CLD-909, like its predecessor, provides the DIN plug receptacle for external computer control, and is priced at under $1000.

Pioneer also produces the LD-V3000, which is functionally equivalent to the LD-V6000 industrial player but offers compactness and portability by limiting itself to playback of 8-inch-diameter videodiscs. It should be noted that, since all videodiscs are read from the center outward, all full-size players can read both 12-inch and 8-inch videodiscs.

Laserdisc Players from Other Manufacturers

Other players capable of Level III operation that are available at modest prices through consumer channels include Magnavox's VC-8040 G4, Sylvania's VP-7400 SL, and TEAC's LV-1000; all of these are functionally equivalent to the Pioneer LD-700. Several Japanese manufacturers are also offering machines equivalent to the CLD-900 that can be controlled in the same fashion as the LD-700.

Unique Features

In addition to the standard features available on all videodisc players, several machines have been designed with unique playback capabilities suited for real-time simulations. These machines can play back *skip-frame* edited discs, where up to five motion segments are edited into alternating frames. With this technique, also known as *interleaving,* the video for two to five different scenes is broken up into frames and recorded onto the disc on an interwoven basis. Then, while playing, the laser continuously skips over the unused frames of video, reading only the picture information for a single scene. If the program needs to branch to the alternate scenes, the laser merely offsets its reading pattern by one to four frames. In this way it is possible to have virtually seamless transitions while switching among several video scenes. Although ideal for action-packed simulations, this technique uses up disc geography very quickly and requires exceptional editing control at the disc premastering stage.

Another feature that can achieve a similar fast-branching result is known as *instant jump.* With this capability, the player can access any frame within 200 frames in either direction of the current location, without noticeably disrupting the display. Players without skip-frame or instant jump will display black on the screen during even the shortest of searches, which interrupts the continuity of the presentation. Motivated by the need for fast-action branching in laserdisc-based arcade games and flight or driving simulators, the designers of the Pioneer LD-V1000 and the Hitachi and Philips disc players incorporated these capabilities. In addition to these features, the Hitachi front-loading VIP-9550 and the VIP-9500 use diode lasers and RS-232 ports. Philips makes the helium-neon laser-based VP-935 and VP-832, both of which feature the instant-jump capability and can read interleaving frames.

LaserVision

All the players mentioned previously adhere to the LaserVision (LV) standard for encoding video, and each can play any LV disc, regardless of manufacturer. LaserVision has become the default standard for interactive video applications because of its performance, durability, and unique playback features. Players that adhere to the LV specifications are also capable of using CLV (constant linear velocity) extended-play videodiscs (60 minutes per disc side), which suit the run-time requirements for movies but lack the frame-access capability and other features needed for interactive use. While RCA's discontinued CED format was not designed for interactivity, other videodisc formats not compatible with LaserVision—McDonnell Douglas's Laserfilm and JVC's capacitance Video High Density (VHD) system, for example—can be used for interactive video.

COMPUTERS AND INTEGRATED SYSTEMS

The user environment and the demands of a particular application generally determine what type of computer will be used to control a Level III system. The availability of interface devices and software development tools will also greatly affect the decision as to which computer to use. The prevalence of the Apple® II family of computers in elementary and high schools seems to make Apple the preferred choice for interactive video in public education. However, the popularity of MS-DOS® machines in the business community points to the use of PC compatibles for that environment. Fortunately, a variety of interfaces and development software exists for both types of computers.

Although Apples and PC compatibles may be the most frequently used, virtually any computer with an RS-232 port can control industrial videodisc players. The high-quality graphics output of the Amiga® and Atari® ST make them suitable options. Sony's SMC-70, well-designed for interactive video applications with its color graphic-overlay capabilities, was somewhat hindered by the 64K limitation of its CP/M® operating system and its late introduction into the microcomputer market. A low-cost interface for the Commodore 64® was marketed by Digital Research, but the value of interactive video was never realized by home computer users. While the Macintosh™ has been used in some interactive applications, its nonstandard video output does not lend itself to use with composite video.

A primary concern in configuring an IVD system is determining how the computer will communicate with the videodisc player. In the case of the Apple II, there are videodisc interfaces that connect to the game port or use proprietary cards in one of the internal slots. For PC-based systems, custom peripheral cards, serial cables, or parallel printer interfaces can be used.

A second question relates to how the computer will accept user input. Some applications will require input from sources other than the keyboard. Touch screens provide an easy-to-use form of access, allowing users to control program flow directly by pointing at information on the display. Light pens are a less expensive way to let users point to their choices, but they are not considered as durable for public use as touch screens. In more specialized Level III applications, it may be desirable to include such input devices as bar code or credit-card readers. With applications that require simulation of real-world experiences, elaborate control and sensor systems can be devised. One such application has laser-emitting police revolvers, which are used to "shoot" at video-display criminals for the purpose of training law enforcement officers in judgmental use of firearms.

In addition to systems pieced together by the user, several companies offer fully integrated packages that include the computer, the videodisc player, the interface, and graphics-overlay hardware. Most are available with development software and peripherals. One advantage of purchasing a system from a single-source vendor is the technical support offered after the sale. It is also generally more efficient to start with a ready-to-run system with all the needed elements than to construct one with components from several sources. Finally, single-source suppliers can sometimes offer better prices because they are packaging more than one component.

Some Currently Available Systems

Sony manufactures and sells the Sony View System, consisting of the SMC-2000 (MS-DOS compatible computer) and the LDP-2000 videodisc player, for around $7500. The SMC-2000 uses a 16-bit microprocessor and 256K of memory that can be expanded to 512K. It was designed for interactive use and can be expanded to include many peripherals. A variety of programming languages and authoring systems are also supported. The View System makes it possible to overlay graphics in 256 colors selected from a palette of 4096 available colors. The combined RGB (red/green/blue) graphics and RGB video permit the use of video projection systems. Options include Level II configuration, digital data capture from videodiscs, and still-frame audio.

NCR's InteracTV-2 system uses a 256K NCR-DOS computer (MS-DOS compatible), expandable to 640K, with a 16-bit microprocessor. The system can control the Pioneer LD-V1000, Hitachi VIP-9550, or Philips 831 videodisc players. It features high-resolution graphics overlay and software control from either NCR Pilot II or Interactive Training Systems' AUTHORITY authoring languages. The NCR computer uses the RS-232 port to command the videodisc player and offers a touch-screen option.

Visage, Inc., of Natick, Massachusetts, offers the V:Station™ 2000. It includes an MS-DOS computer that is capable of driving the most commonly used videodisc players. It will accept Level II program downloads from the videodisc, and it features text and graphics overlay capabilities. The V:Station 2000 allows developers to use several authoring packages for either computer based training or interactive videodisc programs. Visage also offers the V:Station 2080, featuring an IBM AT-compatible computer.

Digital Equipment Corporation (DEC®) makes the Interactive Video Information System (IVIS), which uses a DEC Professional 350 computer and the VDP40 laser videodisc player, a modified version of Sony's LDP-1000A. With optional adapters, it is capable of driving at least two other models from Sony and Pioneer. The DEC Pro-350 computer requires a 10-megabyte hard disk for most applications; courseware is developed using the VAX PRODUCER authoring system.

Digital Equipment's Touchcom II™ delivery system uses a 400K MS-DOS computer with a serial port. Its videodisc player features a 3-second access time, graphics overlay, and touch-screen options. This system has been widely used for point-of-sale, point-of-purchase, information, and training applications.

IBM® announced its entry into the interactive video delivery systems market in June of 1986. The Infowindow Display™ is a graphics overlay and control system that controls both touch-screen and voice synthesis functions on an intelligent terminal linked to an IBM computer and one or two videodisc players. The system itself is expected to be available for $4195 (without the videodisc player, computer, or graphics cards) in the last quarter of 1986, and its accompanying Video/Passage™ software for creating multimedia presentations is priced at $7200.

CONTROL AND GRAPHICS INTERFACES

Level III interactive disc systems require an interface between computer and videodisc player. The interface allows the computer to control videodisc-player functions and, in some cases, the display of the computer and video output. Using standard serial, printer, or computer game ports, several low-cost interfaces give the user complete control of frame access, playback speed, and audio channel selection. More elaborate devices add the ability to display computer and video output on a single monitor by switching between, or by combining, the two signals.

Control Interfaces

As described earlier, most industrial videodisc players use standard RS-232 serial communications for player control. These require no more than a cable like that used with a modem. With serial communication, data is sent to and from the disc player one bit at a time in a specific pattern and at a preset speed (baud rate). The pattern of bits from the computer contains player commands and frame or chapter address numbers, while data coming back from the disc player reports the player's status, including readiness for the next command, current frame location (or frame status), and player modes.

Although there is a trend to standardize on RS-232 serial communications between computers and industrial disc players, most consumer players use interfaces that emulate the signals normally sent out by infrared remote control units. These nonstandard 32-bit serial signals employ a TTL pulse stream that includes two lead-in bytes, a command (or data) byte, and a final delimiter byte.

The important distinctions between the standard and the nonstandard interface approaches are the ease of gaining control from the computer and the type of information sent back to it. While industrial players send back extensive status information, consumer players can communicate only whether they have received a command or if they are ready for another command.

Besides controlling disc-player functions, some interactive videodisc interfaces are also used to manage the video signals. The least expensive of these can alternately switch back and forth between images from the disc player and output from the computer. This technique enables the use of nonstandard computer video—that of the Apple II, for example—without signal processing.

To combine text or graphic output from the computer with images from the disc player, one of several overlay techniques is required. The system used by the Sony SMC-70 and those used by Visage, New Media Graphics, IBM, and others lay RGB computer graphics over the composite video from the disc. This approach has the advantage of retaining the high quality of the RGB graphics, but it tends to visually separate them from the underlying video images. Computers with a standard composite video signal output, such as the Amiga, Atari, and Mindset, use a process known as *genlocking* to mix the two video signals into one. This method allows the use of inexpensive color monitors; however, the image lacks the crispness and resolution of RGB signals. The Sony View System uses a third method that separates the composite signal from the disc into its RGB components and merges them with those from the computer. While retaining the image quality of RGB signals, this procedure blends the images so that the visual separation of the two is not as apparent.

Philips offers the VP935/ISIA (Intelligent Serial Interface Adapter) to control its own line of videodisc players, though the ISIA can also be programmed to emulate other manufacturers' machines. The VP935/ISIA is an RS-232 controller with 28K of programmable memory, addressable by adding EPROMs. The ISIA is capable of reading alternate frames for interleaved data. The VP935/17 is Philips's basic serial control interface. In addition, Philips puts out a parallel controller interface, the VP935/37, with similar capabilities.

Visual Database Systems of Scott's Valley, California, produces a group of control interfaces designed to connect MS-DOS computers to any of the Pioneer line of laserdisc players. Depending on the player used, prices vary from $65 for connections to the LD-700, the CLD-900, and 909 players, to $145 for a parallel interface controller for the LD-V1000. One controller for the LD-V1000 from VDS employs the unique method of taking a serial signal from the computer's RS-232 port and converting it to the videodisc player's parallel protocol. This serial-to-parallel converter/controller sells for $295. Another feature of the VDS interfaces is their ability to control up to four laserdisc players simultaneously. VDS also produces an interface for Level III connections to some consumer digital audio compact-disc players.

Interfaces that switch between computer and videodisc output make it possible to use computer graphics as transitions between video segments while the disc player, during long searches, displays black on the screen. These interfaces also eliminate the need for two monitors. Among the vendors currently marketing signal-switching interfaces are Optical Data Corporation (formerly Video Vision Associates), Allen Communication, Whitney Educational Services, and Systems Impact.

Optical Data Corporation in Florham Park, New Jersey, makes the $140 VAI-II interface to connect Apple II family computers with Pioneer consumer machines.

Allen Communication in Salt Lake City makes a versatile series of control interfaces. The Video-Microcomputer Interface (VMI™) links an Apple to most popular players; it features one-word software commands with communication in serial or parallel modes. Priced at $395, the VMI offers audio and video switching, and will accept software written in either Applesoft BASIC or Apple PASCAL.

Whitney Educational Services of San Mateo, California, offers interface control cards that will connect the Apple, IBM, and Sony SMC-70 to Pioneer laserdisc players, or to Panasonic and Sony VCRs. The PC-500 controller for the IBM sells for $990. Whitney's Supercircuit II for the Apple is priced at $695; the SMC-500-A for Sony's SMC-70 is $895. All of these interfaces allow the user to operate videodisc and videotape players together or separately. Since all of these units can control both tape and disc players, they can be used to design prototype lessons on videotape before committing to the more permanent medium of videodisc.

Another versatile interface is the VID-232 from Systems Impact in Washington, D.C. Its value lies in its ability to handle both parallel and serial connections between Apple, Commodore, or IBM-compatible computers and most videodisc players. It is also designed to be updated for other types of players with the addition of a single chip. Finally, it is addressable by using common ASCII characters from any computer language.

Graphics Interfaces

Although there are situations in which the use of two-screen presentations or display-switching is appropriate, many designers prefer the more costly technique of combining graphics, text, and video onto a single monitor. Overlay text allows the labeling of objects on the screen or the captioning of picture contents. While it is possible to record text directly onto the videodisc, overlay text allows usage of the same video material with changing text or volatile information that changes with time or conditions.

Computer-graphic overlays can be used to highlight areas, to point to locations, or to mask portions of the screen. The use of graphics also affords opportunities to demonstrate abstract concepts without the expense of studio-produced video; also, real-time graphics generation can be used to reflect user input and to save space on the videodisc. One example that demonstrates some of these strategies is Interactive Television's program for the Defense Mapping Agency. This program employs graphic overlays to point out features on video maps and allows military planners to simulate troop movement and strategies. This is quite an advance from the pushpins on pressboard maps used by General Patton.

Video Associates Labs (VAL) of Austin, Texas, makes graphics cards for both IBM and Apple computers. The PC Microkey System™ for the IBM comes in two levels. The Level I PC Microkey System (about $900) works only with videodisc players equipped with external sync, but it will allow the overlay of graphics on video. The Level II PC Microkey System (about $1800) generates NTSC composite video in addition to the RGB output of the Level I. It is capable of producing broadcast-quality graphic signals. For the Apple, VAL's Microkey System offers genlock capability with potentially 2.5 million colors. It sells for $2400 to $2800, including software. For text overlay, a simple character-generator-only card is available for $350.

Online Products Corporation of Germantown, Maryland, makes the GL-512 Color Graphics card with genlock capability or the capacity to switch between video outputs. It is able to display sixteen colors in 4096 hues at one time, or fifteen colors using video overlay with a resolution of 512-by-512 pixels.

Visage, Inc., offers controllers for most current players that work with IBM PCs or compatibles. The V:Link™ 1550 includes RGB overlay capabilities and graphics generation for $2150.

New Media Graphics Corporation in Burlington, Massachusetts, manufactures the PC-Graph-Over® for the IBM PC, XT, and AT for graphics and text overlays. Designed to control the Pioneer LD-V4000 player, it can be adapted with an RS-232 interface to drive other players as well. At $1300 for the complete system, the PC-Graph-Over supports 16-color displays from a palette of 4096 in 640- by 400-pixel resolution. The PC-Graph-Over supports a variety of peripherals, such as touch screen, mouse, graphics tablet, and light pen. For $9850, the GraphOver 9500 offers the same features but can be interfaced with all computers, including mainframes. New Media Graphics also offers NMG-SLIDE™ software to create slide shows, the NMG-PAINT™ electronic paint system for graphics creation, and NMG-FONTS™ to create twenty different fonts with or without drop shadows.

IEV Corporation of Salt Lake City, Utah, manufactures a series of graphic overlay and videodisc controllers. The IEV-10A offers color graphic capability in a 320- by 200-pixel resolution mode for $450. This model will display four colors out of a possible sixteen. The IEV-10A is designed to be used only with the IEV-40, a controller and overlay device that sells for $550. The IEV-60X™, an external subsystem, uses commands sent in either serial or parallel protocols. These control an internal microprocessor to create graphics, overlay them on a video signal, and manage videodisc player functions. The advantage of this approach is that, with the IEV-60X, any computer, regardless of graphics capability, can be used as an interactive videodisc controller. The IEV-60X sells for under $2000.

AUTHORING LANGUAGES AND SYSTEMS

Although it is possible to address almost all laserdisc players directly from the computer through the controlling interface by using computer commands, many interactive program designers prefer to use *authoring tools*. These tools come as authoring languages or as complete authoring system packages.

Authoring Languages

Authoring languages are high-level computer programming languages that have been designed specifically to present instruction, ask questions, analyze user responses, and branch to appropriate follow-up material. Most authoring languages use commands that combine lower-level computer functions into single easy-to-understand command statements. Among these are text and graphic display procedures, algorithms for matching student input against correct or anticipated incorrect answers, and program flow commands, such as jumping to a remedial presentation when a student answers incorrectly.

Authoring languages are intended to make programming of computer-based instruction accessible to nonprogrammers. Even though authoring languages eliminate the terse and technical nature of most computer code, using them still requires some programming skills.

Authoring Systems

In contrast to authoring languages, *authoring systems* eliminate the requirement for a thorough knowledge of command-based programming. Authoring systems use easy-to-read menus, templates, and prompts to allow educators to create computer-based instruction. The systems usually include simple methods for creating lessons and quizzes, and some allow for keeping records on the students' progress. Authoring software for interactive video allows an instructor to present lesson content from video or audio sources on a disc and then, based on a given response from the student, display a still-frame, choose between audio tracks, or search to any location on the videodisc.

While some systems offer a fixed set of instructional strategies into which the developer can insert specific content matter, questions, and answers, others provide a combination of instructional strategy prototypes, along with the capability to create new ones. This more flexible approach, however, again demands the skills of a programmer.

Some Features of Authoring Systems

The features that distinguish authoring systems include the sophistication of the answer-processing algorithms and the ability to import screens or data from other programs. More complete systems provide record-keeping and analysis, as well as support for peripheral devices and the graphics capabilities of different computers. Finally, the most flexible systems enable developers to access "lower-level" programming languages to perform functions not available within the authoring system.

Answer processing in various systems ranges from managing true/false and multiple-choice questions to complex parsers that can interpret freely written responses to open-ended questions. Most packages will accept fill-in-the-blank questions. The weighting of answers and variable tolerances for error are capabilities that suggest thoughtful design in an authoring tool. Systems should be able to accept answers from the user that include imperfect spellings, reordered components, and even partially correct responses.

The much-felt need for training in the use of computer software has led a number of authoring product developers to include the capability to "capture" screens from software displays. Although the value of this feature depends largely on the subject area, it can be used to capture illustrative graphs or data that the authoring system cannot generate.

Record-keeping components are often overlooked in authoring systems, yet they offer significant labor savings for instructors. These features make it possible to manage various data on learning activities, such as item analysis statistics, rank ordering of scores, student progress, and many other measures. Another useful record-keeping function is the ability to insert a place mark at the point where a student must stop an instructional session. Such a device allows the learner to return at any time and pick up where he or she left off.

Support of peripheral devices, in particular support of graphics output on the target delivery system, is of obvious importance for authoring lessons. While some authoring packages provide graphics generation commands, others are limited to loading and displaying graphics files from computer disks.

The primary trade-off one encounters in using authoring tools is that of power versus simplicity. The best authoring systems offer all the needed tools and remain easy to use but still must sacrifice much of the flexibility and power available in their host computer system. This inherent limitation is resolved in some authoring systems by allowing the use of routines or even entire programs written in more powerful languages.

Although authoring software is both bought and sold on the fear of programming, its genuine value lies in its potential for greatly increased productivity in courseware creation. When less specialized training in computer programming is needed, far more educators will better be able to control how content material is turned into instruction.

Currently Available Authoring Languages and Systems

The CDS (Computer Design System) II Authoring Language from Electronic Information Systems (EIS) in Sandy, Utah, is a very high-level language that has been refined since its inception in the early 1970s. While a knowledge of programming is necessary to use the underlying command language CDL, CDS II offers prototype templates that the instructor can merely fill in. Written in the C language, CDS II is designed for use with IBM, Sony, Zenith, Macintosh, and other computers, with just about all of the most commonly used videodisc players. Implementations are available to take advantage of color graphics overlay capabilities from Visage and other overlay hardware vendors. The CDS II instructional creation system is available for around $800, with a delivery-only run-time version—CDSX II—offered for about $300. EIS has also developed a newer package named Genesis, which further simplifies the authoring process by employing a mouse and icons to operate the program.

The AUTHORITY system from Interactive Training Systems (ITS) in Cambridge, Massachusetts, does not require that the user be a programmer. It can be used for both computer-based training (CBT) and for interactive videodisc presentations. Completely menu-driven, it will allow the use of such peripherals as a light pen or touch screen. AUTHORITY works on the IBM PC, XT, or compatible computers and will drive either videotape or videodisc players. AUTHORITY costs $2995 and is capable of designing varied instruction, as well as keeping track of student records with a sophisticated course-management program.

Whitney Educational Services' Insight authoring series provides English-language commands for Apple, IBM, and Sony computers to control either videodisc or videotape players. The Insight PC authoring language translates a designer's English into BASIC commands. With standard record-keeping functions, it sells for $990. The Insight 70 authoring language, which is similar to Insight PC, was designed for the Sony SMC-70 and sells for $895. Whitney's most comprehensive authoring package is Insight 2000 Plus for Apple II computers. The 2000 Plus, which was created for both disc and tape, can randomly access still or motion segments and keep full student records. Priced at $990, Insight 2000 Plus requires AppleWriter™ software.

Laser Write™ from Optical Data Corporation is a simple, easy-to-use authoring language for Apple computers, for which no programming experience is necessary. For $75, the Laser Write language includes its own text editor, and can be used with the AppleWorks™ word processor. This system can create highly structured branching routines, but its answer options are limited. Still, for a very low cost, Laser Write can help an instructor design interesting lessons for interactive video-based training.

Another authoring system of note is QUEST™ from Allen Communication. Designed to run on the IBM and priced at $1295, the QUEST system provides complex branching capabilities, flexible design, and extensive record-keeping functions. QUEST allows "exact," "close," and "phonetic" spelling options from the student/user. The designer can specify the order of words or numbers required for a correct answer, and how critically the answers should be judged. Answers can be weighted for importance, just as on some standard examinations. QUEST features graphics production and touch-screen and light-pen options, and can be used with computers alone, or with computers operating in conjunction with videodisc or videotape.

The field of instructional authoring and the interactive videodisc industry itself were recently joined by IBM with its product announcement of the Video Passage Multimedia Authoring and Presentation System. This software is designed to support the IBM InfoWindow Display hardware described earlier and will be available for $7200, including a reusable run-time program. PA1, as it is called, offers spreadsheet-like authoring with menus, options, and pop-up windows.

The need to reduce the complexity of developing interactive disc applications and computer-based instruction has motivated the development of many more authoring tools than those mentioned here. Frequently, developers faced with unusual instructional problems will create collections of code that speed the process of courseware creation. These tools may eventually be packaged as authoring products, offered as utilities, or simply used in-house. New software packages, or enhancements to old versions, are constantly becoming available as the technology matures and the demands of users grow.

COMPRESSED AUDIO AND DATA

The immense storage capacity of videodisc for images has proven to be both a boon to archivists and a frustration to interactive program developers. Since standard audio uses the two available tracks on a videodisc, it is tied directly to the motion video, which makes it impractical to use in 1/30-of-a-second pieces with the associated frame. While still-frames extend the useful playback to tens, if not hundreds, of hours for a single disc, the lack of accompanying audio unfortunately makes them lifeless. Disc-player manufacturers and others have appreciated the problem and offered a variety of solutions.

The most common solution to the audio-with-stills problem, though neither inexpensive nor simple to produce, has been to store encoded audio in the video bandwidth of frames and use decoders to retrieve and play it back. This strategy yields from 1.2 to 10 seconds of audio per frame, ranging in quality from slightly better than AM radio to telephone-like. This compression nets playback times of 15 to 150 hours per disc side, depending on the technology used.

The simultaneous development of digital audio techniques and general-purpose digital optical data storage technology quite naturally led to the use of videodisc as a data storage medium. Virtually identical to the techniques used for audio-with-stills, the added resource of error-corrected general-purpose data on the same media with images and analog sound is now available to developers.

Sony's approach to adding audio to still-frames, called still-frame-audio (SFA), works with any player that can read automatic picture stops. The audio, originally recorded in an analog signal and transferred to ¾-inch videotape for frame-related editing, is translated into a digital format. It is then encoded as a video signal to be recorded onto the videodisc adjacent to its accompanying still-frame of video. When the disc is played, the digital data is read into a buffer and stored temporarily in a peripheral called the SFD-1. When the still-frame for which the audio is intended is reached, a signal is sent to the SFD-1, which decodes the digital information, playing it back as an analog audio signal through a speaker. Sony's SFA technology is capable of storing up to 40 seconds of audio in 39 video frames (the equivalent of 1.3 seconds of video on disc) in a 256K buffer. It sells for around $2500.

Pioneer's Sound With Stills and Data (SWSD) system operates in a similar fashion. Using digital time-compressed adaptive delta modulation (ADM), the SWSD reduces the bandwidth of the recorded audio signal so that it can be stored in a relatively small amount of space. The audio signal is stored on the videodisc as video information, which can be seen on the screen as snow. Most systems display black while the audio is being transferred to the decoder. The SWSD system uses 10 to 30 frames per audio event. About 9 to 28 seconds of audio can be stored in each frame; obviously, the more seconds used to store the audio, the higher the resultant playback quality, and the more space is required in the video frame. With error correction and three levels of audio quality to choose from, the SWSD sells for $1500.

EECO Inc., of Santa Ana, California, offers the EECODER® VAC 300 Still-frame Decoder for around $2500. The VAC-300 can store up to 10 seconds of audio per still-frame, and each audio event can play back up to 40 seconds of real-time audio. This means that one could store 150 hours of audio per videodisc side if no video were used. The VAC-300 utilizes a compressed analog recording technique, which results in fewer line dropouts than digital recording and decoding methods. EECO's system was used to produce the Medcom Learning Center disc, which contains some 23,000 images and 85 hours of audio on a single side of the disc.

Visual Database Systems makes the SSD-1. This device can be used with Pioneer's SWSD system to let an operator take advantage of the digital data download option of the SWSD system from a laser videodisc, so that data can be loaded from the videodisc into a PC. This is quite different from CD-ROM, as both data and motion video can be mixed on the same videodisc. With a total data capacity of 400 megabytes, the SSD-1 sells for $500.

There are other alternatives to the still-frame audio techniques that are somewhat less expensive and offer much higher fidelity audio reproduction. LaserVideo makes a proprietary interface that connects a digital CD audio player to a Level III IVD system, giving the developer unlimited use of the videodisc while playing very high quality audio from the CD. The LaserVideo approach is not the same as the recently announced Sony and Philips CD-I technology, which can display four resolutions of digital still-image and play four levels of audio fidelity using only the 4.72-inch (12-cm) CD media.

HYBRID CONTROLLERS

Despite the continuous improvement and expansion of IVD technology, some situations call for specialized solutions. In an effort to either reduce the cost or improve the user interface, some companies have developed hybrids of the standard delivery system. These customized systems feature a variety of input devices and combine the programmable IVD control functions normally found only in desktop-style microcomputers.

One such system is the Acorn Touch system from the Acorn Company in New York City, introduced in mid-1986. The Acorn is a self-contained touch-screen monitor unit, capable of storing up to 256K of data, that has its own controller board with programmed EPROMs. The device will also accept Level II program dumps from videodisc. With a resolution of 1024-by-1024 pixels, the touch screen uses a Sony 13-inch monitor and will work with any RS-232–supported laser videodisc player. Acorn Company offers generic programs for its EPROMs, or it will develop programs according to a client's specifications. Since no external computer is required, the $2000 price can be a significant cost savings when a number of delivery systems are needed for a single application.

Seattle's IXION introduced the IXION 100 "micro-center" for videodisc, videotape, videotext, and CD-ROM in mid-1986. The IXION 100 has the ability to control several different videodisc players and emulate other computers using "personality cartridges." It features an 80188 microprocessor with 256K RAM on board. It comes with a variety of user and system interfaces, including an x-y touchpad, two joysticks, two action/enter keys, a numeric keypad, and ten programmable function keys with LED labels that indicate the function that the key is currently programmed to perform. There are two RS-232 ports, one parallel port, input and output for NTSC video, and ports to control a light pen or a mouse. The IXION 100 has seven graphics modes, is overlay-capable, and comes with 512 possible colors and thirty-two sprite planes for animation. It is a "run-only" computer that can decode compressed audio or digital data from a videodisc player and download software from an IBM PC, and that has the ability to decode quality audio with its own music synthesizer chip. The IXION 100, which fits into a briefcase with an 8-inch videodisc player, was designed as a lightweight and portable alternative to bulkier delivery systems. It is expected to sell in quantity for $2000.

NETWORKING

In order to reach beyond the limits of single-user environments, single videodiscs, or the confines of one video display, some developers have produced interactive video networks. By linking several videodisc players to a multi-user computer, a developer can produce an integrated system that serves multiple users from one location. This approach offers the benefit of centralized management and data collection, and can save the cost of redundant computer systems. Online Computer Systems offers the OnLine LAN, which can control up to eight separate videodisc players so that they can share the same courseware. The control software is written in the C programming language and works with the UNIX operating system. Student workstations include a 256K IBM PC with Sony monitor and Pioneer LD-V1000 player, for around $7400. Perhaps the most extensive use of networking has been in IBM's Guided Learning Centers, where individual workstations with videodisc players and terminals operate under the control of a mainframe computer.

Another form of component networking using videodisc players ties multiple players to a matrix of monitors under control of a single computer, creating a video wall. In a fashion reminiscent of multi-image slide presentations, the disc players are synchronized to show the parts of a single image that are combined to form the whole picture, or a collection of related images that together make up a scene. Used primarily as attention-getting props for trade shows, the notion has also been put to use in museum displays.

Visual Database Systems offers a video wall setup with its own videodisc controllers for around $15,000. Another method of achieving the same result is to take a single video source and electronically divide the image, using digital video techniques. The divided sections of the image are then fed to a cluster of monitors to once again form a single whole image when all monitors are viewed in the correct proximity to one another. This type of system runs about $100,000.

Another idea, more like a television network than a computer network, is that of displaying the video information over a television cable system to students and having them respond via a workstation/terminal. Although this method eliminates the expense for multiple computers and videodisc players, it is not far removed from the classroom teaching method of a single teacher working at one pace with a group of students who learn at different levels and speeds.

LEVEL III APPLICATIONS

Once all the video/computing hardware and software is in place, the fun begins: putting all this wonderful equipment to use. Although primarily for education and training, Level III interactive videodisc has gained in popularity as a medium for other uses over the last five to seven years. Currently, it is being effectively utilized for business, the military, psychological evaluation, marketing and merchandising, entertainment, public information distribution, and archival storage of documents and images.

Education and Training

Education and training have probably received the most attention among all Level III applications, and for good reason. With an intelligently designed program from an interactive system controlling the lesson flow, students can learn at the pace most comfortable for them through completely individualized instruction methods. The medium is very popular with students, and statistical evidence shows that IVD training increases retention of lesson material. Teachers can use Level III systems to keep track of students' progress by maintaining records on how much and how fast they are learning. Using a single videodisc, lessons can be "re-purposed" to meet different instructional needs. Interactive videodisc training can be especially effective for training people in widely dispersed geographic areas, or for training large numbers of people who need individualized instruction. It also verifies for the instructor that the material has been learned.

With all these advantages, why hasn't interactive video been more widely accepted in public schools, colleges, and universities? There are a number of reasons, and perhaps the first is that most people are simply not aware that Level III interactive videodisc is an alternative, let alone a viable, instructional medium. Another factor is the inclination in any large institution to stick to tried and true methods. Some educators say they have had enough trouble just trying to get linear videotape into the classroom, let alone interactive videodisc. Some instructors have just begun to become familiar with computers, and they feel that CBT methods using the computer alone are sufficient for their needs. Putting together a complete interactive delivery system can be difficult for those just introduced to IVD; and, more often than not, once they have the system, many teachers become frustrated trying to create lessons using authoring languages that require a knowledge of computer programming. Few complete courseware products with software and videodiscs exist at the moment, although that is beginning to change. Finally, the initial expense involved in setting up a delivery system has proven to be a serious deterrent.

No technology will ever be a panacea for training problems, and Level III IVD should not be viewed as such. Although, for certain types of material, interactive videodisc systems do offer unique learning opportunities that may be superior to other methods for specific applications, usually a combination of methods based on the subject matter and the needs of the intended audience works best. Interactive video is especially appropriate for illustrating concepts that have an inherently visual element. A textbook can describe how to locate a rotor distributor under the hood of a Chevrolet or where to solder a joint on a printed circuit board for a stereo, but Level III interactive video can both describe and show exactly how to do these things.

Currently Available Educational Products

Optical Data Corporation has issued a series of educational videodiscs on the earth and life sciences and astronomy called *The Living Textbook Series*. Educators may purchase the videodiscs as part of a package that includes an instructor's guide with floppy diskettes, the Laser Write™ authoring system, the VAI II control interface, an Apple keypad, and a consumer laserdisc player. All components, including the discs, may also be purchased separately. These science archives include thousands of still and motion images of our earth and the planets, the space program, and the life sciences; in addition, suggestions for lesson plans accompany the videodisc and its guide.

VideoDiscovery, of Seattle, created the *Bio Sci* videodisc, which includes more than 6000 high-resolution biological images and a cross-referenced subject and item index for quick access to the pictures relevant for a particular lesson. VideoDiscovery has also produced menu-driven floppy disks containing information on each image to help explain the pictures as they are displayed. Some of the topics covered are entomology, ichthyology, ecology, cell biology, and biochemistry.

Several other interesting interactive educational progams have been created, including those from Systems Impact, SVE, and EduTech. *The Delaware Music Series, The Business Disc,* and *The Whale Disc* from National Geographic are only a few examples of the current selection.

Job Skill Training

One application for Level III programs that has received quite a bit of attention has been the use of interactive video for employee and management training. An obvious use for this application is in job skill training. Employees who work with technical products or subject matter may find that learning some aspects of their job from an interactive visual presentation helps them adapt faster than if they were learning the material from a book or lecture. This is especially true if the topic requires visual recognition, as in parts manufacture or mechanical repair, or if certain ideas might best be handled with illustrative graphics or animation to communicate concepts too complicated to explain with words alone. Deltak Training offers a series of technical skills training discs that are in experimental use in a number of community colleges.

Interpersonal and Management Skills

Another area where employee/management training has benefited from the use of Level III IVD is the field of interpersonal skills and attitude awareness education. Wilson Learning's Interactive Technology Group used Sony's View System to reduce a 2½-day management skills seminar into an 8- to 12-hour course on seven interactive videodiscs. The course is designed to help managers understand why they easily relate to certain types of people and how to develop good relations with other types. In one lesson, for example, using humorous examples, four different personality types on the screen interact in a role-playing office situation. The viewer then answers questions categorizing these people according to such characteristics as "pushy," "stuffy," or "nice."

Alamo Learning Systems has produced the *ALS Process Management Skills Program* for such clients as General Motors and General Electric. Using the Sony View System, one section of the management skills program places the viewer in the position of solving problems that might be faced by a manager. The situation is presented in the form of a "business decision-making mystery"—the viewer assumes the role of manager and makes decisions that affect the company and employees. Depending on the "manager's" answers, one of five possible endings is shown in motion video. Thus, by actually making decisions and viewing the examples of others, managers can learn how to handle major and minor company crises.

Digital Equipment Corporation offers *Decision Point,* a program for business that is designed to be used with DEC's IVIS hardware. In one case study, the viewer assumes the role of Associate Vice President of Sales and makes decisions that, it is hoped, will result in increased sales. Again, humor is used to make important points, and the system keeps track of the VP's choices. Using a gamelike format, the system informs viewers about the effectiveness of their decisions.

Military

The United States military, which is in the process of converting to an Electronic Information Delivery System (EIDS), is one of the largest users of Level III interactive videodiscs. Currently, the Army has requested bids for 30,582 IVD systems, and the National Guard has asked for another 9976. Systems delivery from the hardware manufacturers is required within 300 days after the contracts are awarded in September 1986. It is estimated that the military will order between 15,000 and 50,000 more systems for training purposes during the period from 1987 to 1990.

The military specifications for such a system include an MS-DOS computer with a 16-bit microprocessor, an RS-232 port, two 3.5-inch disk drives, two unused expansion slots, and 256K memory, expandable to 512K. Also specified is a front-loading videodisc player with a 3-second maximum search time and a 100-frame quick-jump capability. In addition, the player should be able to handle still-frame audio (10 seconds in no more than five frames) and digital data program dumps from the videodisc. For some designers, these specifications outline the ideal wish list for the perfect IVD delivery system.

The Level III videodisc medium is not exactly new to the military—most branches have been using it in training for several years. The Army has specified that most of its linear training tapes be transferred to videodisc with appropriate supplementary materials. Several sophisticated flight simulators have been produced using videodisc combined with real-time computer-graphic animation. The U.S. Army Signal Center at Fort Gordon, Georgia, produced an extensive series on using electronic communications equipment. It reports that the Center expects to generate anywhere from 120 to 140 more disc sides in 1987. Some of the most impressive IVD programs are those from the Defense Language Institute. These are designed to teach military and diplomatic personnel how to speak foreign languages using real-life situations videotaped in the destination country. Computer graphics, text, and complicated branching not only teach the language but also make the viewer aware of each country's customs.

Health and Medicine

Educators in the health and medical fields are also taking advantage of the unique potential of interactive videodisc for both training and archival storage of medical slides. In Minneapolis, Health EduTech has produced an informative disc on acquired immune deficiency syndrome (AIDS) for medical professionals and the general public. Many hospitals have expressed an interest in IVD training, and medical schools have also begun to utilize the medium.

Marketing and Merchandising

Marketing and merchandising is another area in which IVD is just beginning to expand. Already several of the larger retail chain stores have used point-of-sale (POS) and point-of-purchase (POP) videodisc kiosks to increase their sales. With a *point-of-sale* system, customers may be presented with a visual catalog of items, which they can examine as long as they wish. The catalog information helps customers make better purchasing decisions and frees salespeople's time for other tasks. The *point-of-purchase* kiosk is an IVD setup allowing the customer to use a credit-card reader to buy any given item and have it delivered either in-store or by mail. ByVideo of Sunnyvale, California, offers such a system. Customers can select Florsheim shoes, specifying the size and color, and then pay for the shoes by credit card and have them shipped to their home.

Another type of POP system is what is known as the *infomercial.* An attention-getting video piece, called a "grabber," plays repeatedly until a viewer is enticed to touch the screen or a key, or until an ultrasonic sensor detects potential users gathered around the display. Once the screen, keyboard, or sensor tells the videodisc system that it has an audience, it will usually display an animated video segment designed to maintain even the shortest attention span for a few moments longer. Then the program offers subtle, seductive commercial messages. Some of the more advanced kiosks can estimate the number of people standing nearby, so that marketing people can draw conclusions about the number of people reached using that medium versus the actual number of sales of a particular item.

Point-of-sale kiosks have been used at banks. They have been used to sell soft drinks, clothing, cosmetics, and numerous other consumer items, and their use is expected to increase dramatically over the next few years. Point-of-purchase applications have been more limited in their penetration of the market and are, at the moment, still in the experimental stages. The "infomercial" videodisc kiosks have been well received so far, though more study is needed to know exactly how much they affect consumer buying habits.

Games

Most people probably first came to know about videodiscs by playing a once-popular videodisc arcade game called *Dragon's Lair,* the product of former Disney animator Don Bluth. For fifty cents, the game player uses a joystick to make split-second decisions to guide the hero (Dirk the Daring) through a maze of perilous decision points to rescue Princess Daphne from a fierce dragon. By moving the joystick at the proper moment, the videodisc player jumps to an action sequence that shows the consequence of the player's move. If the move is made at the wrong time, the player watches as Dirk dies a horrible death; if all the proper moves are made within a preset time-window, Dirk wins a kiss from the Princess.

Several other videodisc-based arcade games—*Space Ace* (also from Bluth), *Firefox,* and *MACH 3*—have come and gone since *Dragon's Lair.* Even though video game arcades have lost their appeal to the public, the initial push for videodisc players in the game arcades advanced the technology with instant jump and interleaving, and improved the sturdiness of videodisc players. (It was because viewers objected to seeing a few seconds of black on the screen when the game arrived at a branching point that designers tried switching between two players to eliminate any pause in the video and, subsequently, that the arcade-game distributors helped make the case for interleaving and for stretching the limits of the instant-jump capability of the disc players.)

For the home market, most videodisc games have relied on Level I applications, such as *The First National Kidisc* and *Murder Anyone,* that could be easily accessed with nothing more complicated than a random-access remote control.

RDI Video Systems produced Halcyon, the first integrated IVD system for home use, which used voice recognition and speech synthesis for a Level III game entitled *Thayer's Quest*. Priced at $2300, Halcyon was a fascinating attempt to bring Level III into the home, but its audience was limited to the well-heeled curious and the video-obsessed.

Films

When producing films on videodisc for the home market, most manufacturers choose the CLV format, since it can hold up to 60 minutes of video per side. One exception is Criterion Collection of Los Angeles, which has issued films in the CAV format. This means that videophiles and film buffs can access the still-frame, slow-motion, and scan features of most players to better examine the films. It also indicates the possibility of Level III applications. Criterion's *Citizen Kane*, for example, comes on four disc sides with a second audio track explaining the significance of a shot or sequence, as well as how it was produced. After the film, a series of still-frames details the conception, writing, making, advertising, and critical scorn and acclaim heaped on Orson Welles's classic.

Music

Some videodisc producers thought music videos might be the way to the pocketbooks of young audiences with disposable income, but it appears that the day has yet to arrive. A few experiments with videodisc jukeboxes have been tried in some large cities; for a few quarters, a viewer can select one of forty-eight CLV music video titles played on one of three players inside the jukebox. Although videodisc jukeboxes have done well in some international locations, most have met with limited success at best. An interesting experimental variation on this idea has become quite profitable, however. Miami's Video Jukebox allows cable TV viewers to call in and request their favorite music videos from a similar videodisc playback setup. The program lists a 976-number, and viewers can simply dial in the number of the music video they wish to see. This venture has been so successful that it is now able to sell advertising for what was originally a noncommercial cable experiment.

Public Information Distribution

Another field where videodisc is finding a comfortable niche for itself is that of providing information for museum or exposition visitors. Disneyworld's Experimental Prototype Community of Tomorrow (EPCOT) uses videodiscs with varying degrees of user interactivity to tell visitors about advances in modern technology and to help them locate sites of interest. The San Francisco Zoo's Primate Center has a Level III presentation about apes, which is designed to be operated by children. At EXPO 86 in Vancouver, British Columbia, IBM has installed its InfoWindow Display system in IBM EXPO INFO kiosks all over the EXPO grounds. The InfoWindow system describes events, pavilions, and IBM products.

Technology For People (TFP) has produced an interesting series of Cityguide videodisc kiosks, designed to help newcomers find their way around. Programs such as *Sarasota at Your Fingertips* in Florida and *The Capital at Your Fingertips* in Ottawa, Canada, provide videodisc-based information centers with tips on finding landmarks and attractions; maps; guides to culture, cuisine, and entertainment; and other useful facts for tourists. The Ottawa disc has two audio tracks, one in English and one in French. TFP's Motourist Info Center provides a similar service, detailing how to obtain food, shelter, and entertainment at videodisc kiosks located in gas stations, restaurants, and truck stops. Currently, the Motourist discs are on display in Ohio, Michigan, and Indiana. These kiosks use advertising to help defray costs, and they issue printed discount coupons for tourists. A similar system, called *The Paris on Videodisc City Guide*, has been set up in France by ADIC Presse.

Information Storage

An application that makes use of the videodisc's prodigious storage capacity is that of information storage and retrieval. Since a CAV disc can hold up to 54,000 high-resolution still images per side, it is possible, for example, to store an entire encyclopedia on one videodisc side. KnowledgeSet has produced the *Knowledge Disc*, which contains the entire Grolier's *Academic American Encyclopedia* on just one disc side. The *Knowledge Disc* is a Level I disc, cross-referenced and indexed for easy access to any subject. Experiments with Level III indexing of encyclopedias are currently under way—CEL Communications' $10,000 *Video Encyclopedia of the 20th Century*, for example.

Librarians and scholars have begun to realize the savings of money and space that can be achieved by using videodiscs as a medium for storing large amounts of visual information. Furthermore, CAV disc storage means that access to information is much faster and easier than wading through stacks of Dewey decimal cards or microfilms. Visual Database Systems (VDS) put together systems for accessing three archival videodiscs in the National Air and Space Museum, detailing in still photographs the history of aviation and space exploration in the United States. VDS used a video information retrieval system based on a Level III setup that was combined with a dBASE® II program to make it easy to locate the precise image desired in just a few seconds. The Library of Congress is currently storing photographs and other fragile images on videodisc, which will enable scholars to easily find and study hard-to-handle historical material. Videodisc Publishing has produced *The National Gallery of Art* disc, an archive of the gallery's extensive collection. Also, Philips commissioned one of the more interesting videodiscs of fine art material with its retrospective of the works of Vincent van Gogh.

In England, one of the largest archival uses of videodisc is the Domesday project—a survey of the British people and countryside that is intended for both mapping of the landscape and a cultural time capsule of Britain. The project will use more than 2 million pages of information, 20,000 survey maps, and 120,000 photographs. *The Videodisc Monitor* in 1985 referred to the Domesday project as the "single largest and most ambitious videodisc project ever undertaken."

Maps

As mentioned earlier in this chapter, the U.S. Army and the Defense Mapping Agency have been working with the Interactive Television Company to create some thirty-nine interactive disc maps of the United States, the European theater, and Korea. These are computer-controlled videodisc-based systems that rely heavily on overlay graphics for dramatizing a given military situation, allowing user input for access and control. Other versions of videodisc maps have been created for police, fire, medical, planning, and other agencies. The U.S. Geological Survey has collected numerous satellite photographs and categorized them on videodisc with computer database programs to easily reference information on crops, land temperatures, climate, and surface topology. One interesting experiment with videodisc mapping was done with the Aspen movie map, produced by MIT's Machine Architecture Group in the early 1980s. With a Level III system, the viewer could "drive" around Aspen, Colorado, using a joystick to control direction. With the press of a key, the user could view the same location during each of the four seasons.

CONCLUSION

It is apparent that the future of Level III applications of interactive videodiscs is a promising one, with prospects for continued growth and usage. It is entirely conceivable that, within ten years, many people will consider videodiscs just another necessary peripheral to the computer. Whether or not the videodisc medium catches on with the home market in the foreseeable future, it has a bright, unlimited potential in many other application areas.

6

CD-I: The Medium of the Future

Larry D. Lowe

__Larry D. Lowe__ is a veteran interactive videodisc systems designer based in San Diego, California. He designed Level III interactive videodisc systems for the 1982 and 1984 World's Fairs. The system for the 1984 Fair used a custom hardware responder subsystem to allow a videodisc-based "persona" to interact with up to forty-eight users at a time, as part of the interactive videodisc classroom of the future. This system was subsequently acquired by the American Museum of Natural History, Smithsonian Institution, Washington, D.C.

Mr. Lowe is currently a consultant to Philips Home Interactive Systems on the development of a CD-I applications developers' guide, and is working with American Interactive Media (AIM) on the design of an initial demonstration CD-I disc. In his spare time, he is a "semiretired" airshow pilot.

In March 1986, at the First Microsoft CD-ROM Conference, David Geest and Richard Bruno of Philips announced the joint intention of Philips and Sony to submit specifications for a new medium based on CD-DA (compact disc–digital audio) and CD-ROM (compact disc–read only memory). This announcement was perhaps the ultimate "good news/bad news" joke for the interactive video systems designer.

The new medium, dubbed compact disc–interactive, or CD-I, represents a complete fusion of video, audio, text/data, and software storage and retrieval processes onto one medium with one format worldwide. The good news is that it is a comprehensive, well-thought-out standard that will be developed and licensed by Philips in the same fashion that CD-DA was, thus ensuring a wide variety of players with a compatible set of "base case" features and hardware manufacturer-specific extensions. On one ergonomic, economic digital disc, the video systems designer can store multiple quality levels of both video and graphics, multiple quality levels of audio (up to around 20 hours), copious quantities of text, ASCII or binary data, and computer software in its most potent form—machine-executable 68000 object code.

The bad news is that CD-I is not here yet—at least not as of the summer of 1986. However, the important point for interactive video designers to realize is that the international standardization of CD-I is a lengthy and complex process, involving several steps leading to a worldwide capability of any player to play any disc. Philips and Sony have spent a lot of time developing a preliminary set of specifications known as the *Provisional Green Book,* which will soon be made available to those hardware manufacturers who propose to develop and produce CD-I compatible players. After a suitable analysis period, the manufacturers will respond with their concerns. After those concerns have been studied by the standards committee, a final revision of the *Green Book* will be published. That point will probably occur in early 1987, and only then can hardware development and authoring-tool creation begin in earnest. Thus, it will be mid-1987 or later before large scale CD-I software development can commence.

I am telling you all of this somewhat gloomy news because CD-I is too significant an advancement to be ignored. The microcomputer industry, in particular, has seen its share of product announcements that were not supported by product, and the casual reaction to Philips's announcement might be taken to mean that the announcement was yet another marketing ploy. However, Philips was careful to make clear in the wording of its announcement that it is presently in the process only of standardizing the specifications.

The standardization of CD-I is indeed real. Many years of effort have already been put into the project; however, many more will be put in before the first CD-I disc is played on the first CD-I player. So what can an interactive systems designer do while the standardization process is still under way? How do you start to build an egg if the chicken isn't available for study?

This chapter provides information that an interactive video systems designer will need in order to develop a valid CD-I concept and begin design work. It also points out the problems and issues inherent in a medium that blends so many traditional disciplines. It will address five areas of information that will give the interactive video systems designer a starting point in CD-I.

First, there will be a general discussion of CD-I technology that will define the scope and nature of the medium, as well as define some terms and conventions. Second, an overview of the information released so far about the preliminary specifications will provide "hard" facts about the standard as it exists today. Third, a design-issues discussion will attempt to define genres of CD-I applications, provide the designer with design insights and conventions, and define the modes of a successful consumer CD-I application. Fourth, a discussion of methodology will outline the processes involved in production of a design. And, finally, a section on publishing will discuss ways that a design might reach the marketplace.

A GENERAL DISCUSSION OF THE CD-I STANDARD

Efforts to describe CD-I are a bit like the old tale about the seven blind men and the elephant. Each man was asked to describe the elephant based on what he could feel. Each came up with a different description, but none developed a true understanding of the beast. Since no previous single system has approached the level of functional integration of CD-I, we are all, in a sense, blind men.

CD-I Is What You Make It

If you are a videodisc designer or producer, CD-I is a specification for a pocket digital videodisc player. If you are a retailer or a stock analyst, CD-I is a logical consumer upgrade from CD-DA. If you are an educator, or a publishing firm that specializes in textbooks and/or filmstrips, CD-I is a fully programmable electronic filmstrip projector. If you are a personal-computer user, CD-I is a terrific personal computer. If you are a video artist, then a CD-I box, along with the proper software, is a personal video-editing/production system that will allow you to develop your own graphics and video effects and overlay them on video from a storage device, such as videotape or disc. If you are a computer manufacturer or original equipment manufacturer (OEM), CD-I is easily the winner of the outstanding new peripheral award. If you are a lawyer, a medical researcher, or a student, a CD-I player and one or two discs will put entire libraries at your fingertips. For that matter, you can bet that CD-I will revolutionize information research and retrieval in the home, as well.

But regardless of your expectations for CD-I in the future, one thing is clear in the present: Philips and Sony believe that CD-I is the medium that will successfully mass-market interactivity to the electronics consumer. While there are many possible applications of the CD-I standard in industrial, educational, point-of-purchase, and other markets, *the object of CD-I is to set the standard for the generic home information and entertainment system of the future.* Since the consumer market is the main thrust, applications designed for that market will be used as examples in this chapter. The many other markets and applications of CD-I will involve a subset of principles presented herein regarding consumer information and entertainment applications.

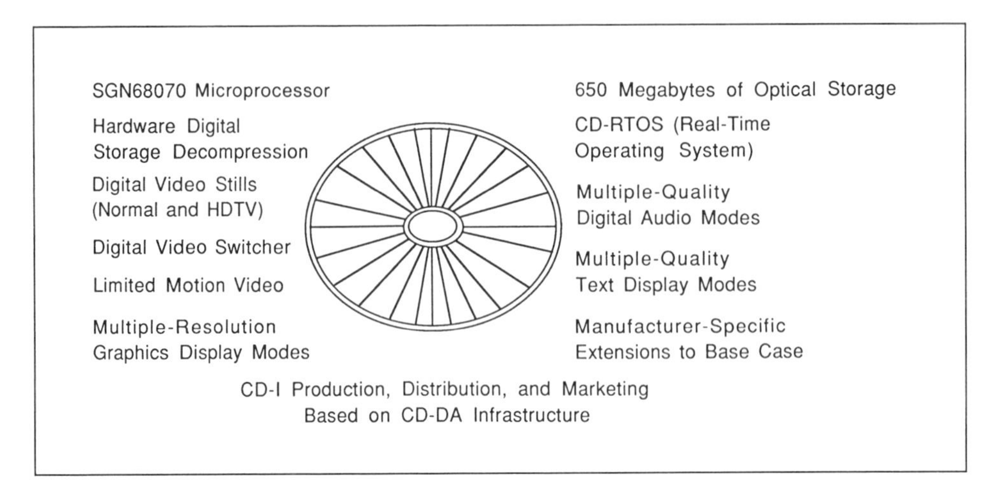

Fig. 6-1. The components of CD-I.

Some Implications of the CD-I Standard

The basic fusion of technology embodied in the CD-I standard is that of advanced microprocessor technology and the very large digital storage capabilities inherent in the optical storage media (see Fig. 6-1). The Motorola 68000-family microprocessors are very powerful and popular general-purpose chips that can be found in most graphics-based personal computers, including the Macintosh, Amiga, and Atari, as well as the more powerful workstations such as the Sun, Sage, and many others.

The CD-DA and CD-ROM standards involve media and player specifications for two forms of digital storage. The first is dedicated to the reproduction of music. The second is a general-purpose digital data storage format with enhanced error detection and correction required for storage and retrieval of advanced data types, including computer program code.

Thus CD-I, which specifies both the microprocessor and the storage format of various kinds of information on an optical disc that makes use of the CD-DA form factor and drive mechanism, is an entirely digital medium and, as any computer programmer will tell you, therefore inherently a very versatile medium. The first conceptual leap an interactive video systems designer must make is the realization that all information is in digital form, and therefore it can be manipulated by the microprocessor once it has been retrieved from the disc. One of the implications of that notion is the concept of storing the basic components of a frame of video, plus the editing instructions, on the disc and letting the hardware do the editing at playback time, rather than doing the editing in the studio and storing the "canned edited" sequences in video.

The remaining aspects of the standard have to do with the formats for the storage of various quality levels of video, graphics, audio, and text information on a CD-I compatible disc and with the hardware required to retrieve and display them. The specifications provide for a number of formats of each kind of information, and it is the designer's job to select the best format for a particular kind of communication task.

Unlike analog videodisc, there is only one stream of information coming from the optical disc, and that stream comes at a relatively slow rate. The only way a designer can switch audio channels—to change languages for example—

is to interleave the data for each channel into one stream and select only the components out of the stream that correspond to the channel desired. The added design effort needed to understand the implications of interleaving, as well as the relationship of data volumes for various kinds of information (ranging from very high for hi-res video, to very low for text and program code), is offset by the additional flexibility offered by the use of up to sixteen audio channels.

The important consideration in selecting a mode of either audio or video is the information density of that mode in a single digital information stream. The basic unit of information is about 2K (2048 bytes) long, and the way to visualize the density of an audio mode is to consider a set of sixteen blocks as the maximum unit. For example, full-quality CD-DA will require all sixteen blocks, leaving no room for any other information. A single mono channel of AM-quality speech, on the other hand, will require only one of the sixteen blocks, leaving the remaining fifteen blocks available for other information, such as video, graphics, or text.

The refresh rate of a given graphic or video mode will depend both on the amount of information it takes to build that video or graphic and on how much audio information must be interleaved with it for real-time audio/video coordination.

Another important concept is that of the *base-case* standardization of the hardware. As in CD-DA, Philips will license hardware manufacturers to produce CD-I players on the condition that their players conform to a minimum set of performance and capabilities standards known as the "base case." This will ensure that any disc mastered to the specifications of the *Green Book* will play on any player that has a CD-I logo on the outside. The standardization of hardware on an international level is one of the most exciting aspects of the standardization of CD-I. There is no constraint, however, on hardware manufacturers past the base case, and the CD-I software community will eventually see extensions to the base case that will include very powerful personal/business computer systems; portable education/entertainment systems; systems specifically designed for installation in cars, boats, and aircraft; and CD-I systems that are built right into the electrical nervous system of the home. These extensions will provide both new programming opportunities and new challenges.

Marketing and Manufacturing Considerations

One aspect of all of this not presented in the specification is the important marketing and manufacturing consideration that CD-I is based on CD-DA. Compact disc–digital audio is an overwhelming consumer electronic success story. And compact disc–interactive can take immediate advantage of those discs sold, the factories that make the discs and players, the marketing and distribution channels that sell the discs and players, and, finally, the customers who buy them.

Seldom has a new consumer technology been so spring-loaded for success. CD-I already has in place a well-developed production, marketing, and distribution infrastructure that will be up and running on the day that the first hardware and software are released. And that CD-DA background support is destined to merge with the LaserVision (LV) videodisc format. Pioneer already makes a

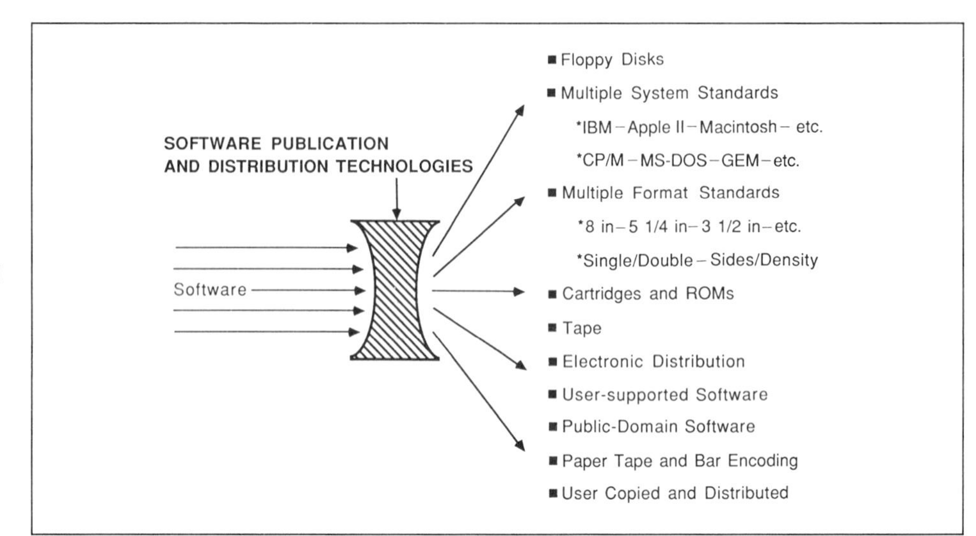

Fig. 6-2. The many ways of publishing and distributing software.

player that plays LaserVision videodiscs and also decodes the digital-audio stereo sound track, if one is present. It is only a matter of time before we see a CD-I/LV player that plays a 12-inch CD-I/LV videodisc. That product is already being referred to as the "Omni Player" in CD-I press releases.

The Potential of CD-I as a Focal Technology

As for where CD-I applications will go once everyone gets through transferring their existing applications into a CD-I format, that is anyone's guess. The fusion of all of this technology results in an enormous potential for heretofore impossible applications.

One way of looking at the potential is a concept put forward by Dr. Bernard Luskin, president of American Interactive Media (AIM), a Santa Monica firm chartered to nurture program development for CD-I. Dr. Luskin calls CD-I a "focal technology"—one in which several major communications technologies become focused onto one medium, as opposed to more conventional publishing technologies, where the information is often dispersed via disparate means.

If you look at text publishing, for example, you will see that depending on the content and intended market, there are many ways to publish and distribute text information. And nowhere is the diffusion of an information stream so apparent as in the computer software industry (see Fig. 6-2). Here, the proliferation of technologies means that a product can be produced and distributed in any of several, often incompatible, ways. In fact, the information stream becomes fragmented *because of* the differences in hardware and operating systems.

CD-I, on the other hand, focuses several information and entertainment publishing industries on one medium, with one international standard for production and distribution (see Fig. 6-3). With a single medium and one worldwide set of publishing requirements, publishers can address an extremely large group of consumers by creating *one* version of their product.

In the future, the distinctions will blur to the point that, when you buy a CD-I disc, you can expect a blend of text, audio, video, and software, with

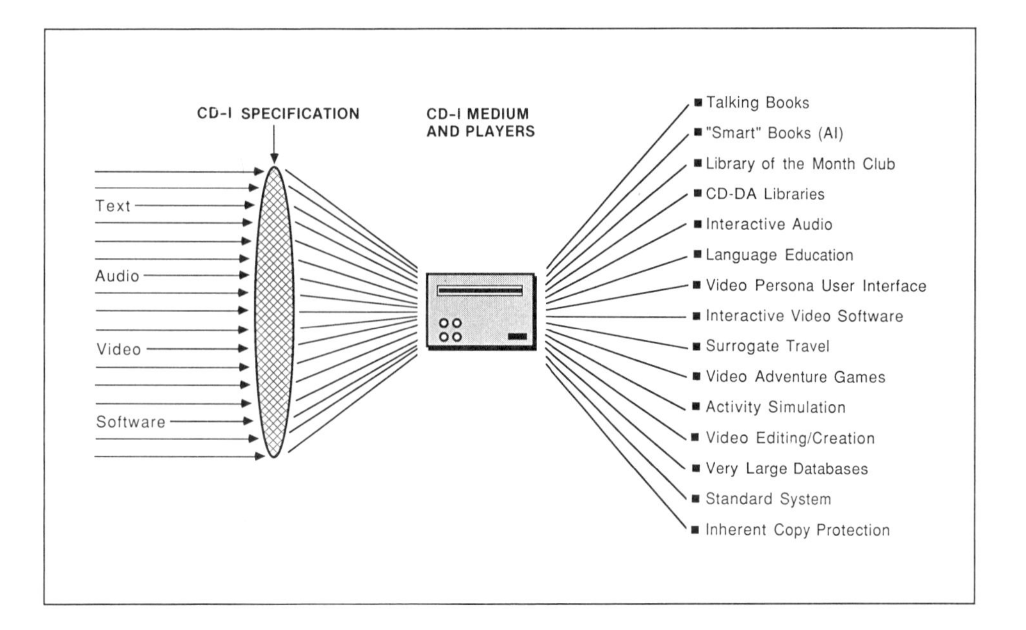

Fig. 6-3. CD-I: A focal technology.

several levels of involvement ranging from an automatic display mode for passive armchair viewing to highly interactive experiences.

A TECHNICAL OVERVIEW OF THE CD-I SYSTEM

First and foremost, CD-I is a stand-alone personal computer. At the heart of the CD-I specification are programs in 68000-compatible object code that will be stored on the disc and used to control the application (see Fig. 6-4). The rationale here is that machine-executable code is required to run applications rapidly and in real time.

Philips intends to use the Signetics PCB68070 microprocessor in its CD-I systems. The 68070, an advanced development of the Motorola 68000 chip, is ideally suited for the task, since it includes two channels of direct memory access (DMA), which allow for data transfer directly from a device to memory. The

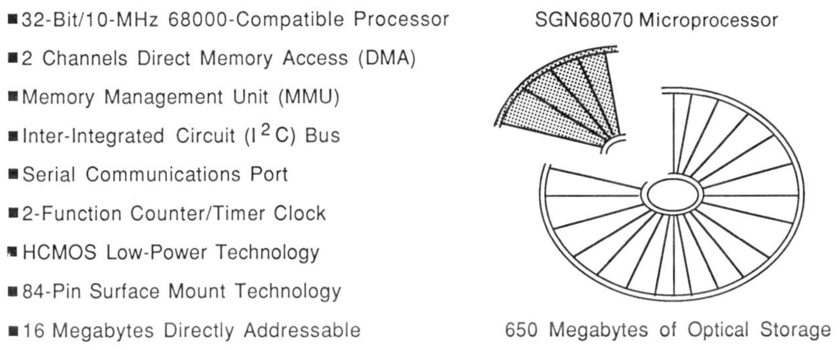

- 32-Bit/10-MHz 68000-Compatible Processor
- 2 Channels Direct Memory Access (DMA)
- Memory Management Unit (MMU)
- Inter-Integrated Circuit (I^2C) Bus
- Serial Communications Port
- 2-Function Counter/Timer Clock
- HCMOS Low-Power Technology
- 84-Pin Surface Mount Technology
- 16 Megabytes Directly Addressable

SGN68070 Microprocessor

650 Megabytes of Optical Storage

Fig. 6-4. The heart of CD-I.

use of DMA significantly reduces the time that the processor spends loading memory from external devices and allows other processing to occur while such data transfers take place.

All of this means that interleaved CD-I video and audio data stored on a disc can be loaded into the video and audio buffers by the two DMA channels, leaving the 68070 free to run the application program. This will eliminate the pauses that currently occur in an audio segment while a new graphics background is loaded into memory.

Also on board the 68070 are: a subset of the standard memory management unit (MMU) chips for extended virtual memory management, an inter-integrated circuit communications bus (useful for coordinating the efforts of a number of high-powered parts on the same printed circuit board), a two-function counter timer, and a serial communications port. (A 68070 chip replaces about $100 worth of silicon in a design.)

In a CD-I player, the power of the 68070 will be harnessed by CD-RTOS (compact disc–real-time operating system), a powerful modular operating system based on OS-9. CD-RTOS can modify itself to handle any new system configuration or special requirement on the fly, so an application can add operating system extensions as needed. In this fashion, the distinction between the operating system and the application tends to blur, and the requirement for the applications programmer to thoroughly understand the operating system is increased.

In addition to the 68000, the CD-I specification includes a lot of custom-developed silicon to do such things as video and graphics decompression, which allows the player to store a 108K graphic frame in 10K on disc and to perform digital switcher effects, such as wipes and other transitions. It is hoped by the designers that the DMA circuitry and the video compression hardware will offset the inherently slow data transfer rate of the CD-I drive, which is left over from the CD-DA standard.

Video

The CD-I specification calls for two video modes and several graphics modes (see Fig. 6-5). Starting at the top, a CD-I player will be capable of limited full-motion video—in this case, limited to a window one-third by one-third of the screen in size, and running at less than 30 frames a second. This limitation is a reflection of the fact that the video is entirely digital, and the data transfer rate from the drive is the same as CD-DA.

Next is the digital YUV coding process that stores a full-screen video image in 328K of memory noninterlaced, or 650K of memory interlaced. The video compression scheme reduces this to 108K of data stored in coded form on disc, and the decoding or decompression is done in hardware at the time the video still is retrieved.

There are two modes for storing graphics information on a CD-I disc. The first, red-green-blue (RGB) encoding, stores the graphics information for RGB display directly on disc with either 15-bit color resolution using 215K of storage to provide 32,768 colors, or 8-bit color resolution using 108K of storage to provide 256 colors. RGB encoding has the advantage that the user can manipulate the image after it is loaded into memory by means of a paint program, or

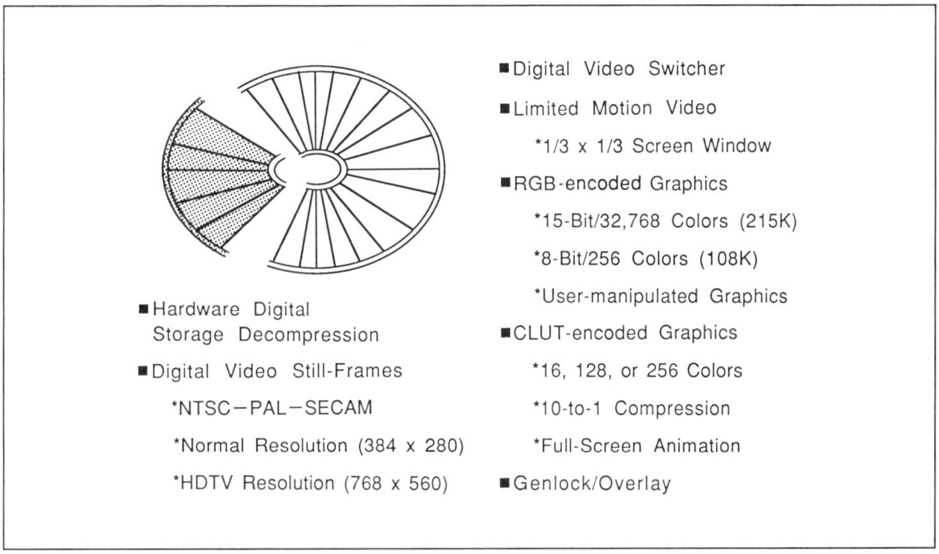

Fig. 6-5. CD-I video/graphics.

- Hardware Digital Storage Decompression
- Digital Video Still-Frames
  - *NTSC—PAL—SECAM
  - *Normal Resolution (384 x 280)
  - *HDTV Resolution (768 x 560)
- Digital Video Switcher
- Limited Motion Video
  - *1/3 x 1/3 Screen Window
- RGB-encoded Graphics
  - *15-Bit/32,768 Colors (215K)
  - *8-Bit/256 Colors (108K)
  - *User-manipulated Graphics
- CLUT-encoded Graphics
  - *16, 128, or 256 Colors
  - *10-to-1 Compression
  - *Full-Screen Animation
- Genlock/Overlay

the user can cut and paste parts of several images stored on disc to create a composite image for output to tape.

The second graphics mode makes use of a color lookup table (CLUT) to encode pictures in three color resolutions. The highest is 256 colors, using 8 out of 24 bits, and requires 108K of display storage per picture. CLUT graphics can be compressed to about 10K per picture for storage on the disc. This means that using CLUT graphics, full-screen animation is possible by simply reading the data off the disc and refreshing the entire screen, as opposed to using the micro to calculate and adjust small portions of the image. It should be noted that CLUT graphics are not intended for use in applications where the user will manipulate the images.

Audio

Audio can be stored on the CD-I disc in seven modes, at four quality levels (see Fig. 6-6). Since CD-I is based on CD-DA, it can play any CD-DA disc or short segments of a CD-I disc at that very high quality level. The CD-DA format allows for a maximum of 72 minutes of stereo audio, and the quality of the sound reproduction is outstanding.

A slightly lower quality music mode is hi-fi, which is almost equivalent to LP music quality and doubles the playing time available. Don't let the LP-quality tag fool you. In a recent test, twenty audio experts listened to samples of both modes and only one could tell the difference. There are two ways to use the hi-fi music mode—stereo or mono, where different audio information can be presented on separate tracks. Mono mode, again, effectively doubles the amount of audio storage available, resulting in four times 72 minutes, or 288 minutes (4.8 hours) of playing time.

Next, there is a mid-fi music mode (great for sound effects in games), where 4.8 hours of stereo or 9.6 hours of mono can be stored. Mid-fi music mode is equivalent in quality to FM broadcast.

■ CD-DA Mode (Existing Standard)
- *1 Track/72 Minutes/Stereo
- *16-Bit PCM Encoding

■ Hi-Fi Music Mode
- *2 Tracks/2.4 Hours/Stereo
- *4 Tracks/4.8 Hours/Mono
- *8-Bit Adaptive Delta PCM Encoding
- *LP Music Quality

■ Speech Mode
- *8 Tracks/9.6 Hours/Stereo
- *16 Tracks/19.2 Hours/Mono
- *4-Bit ADPCM Encoding
- *AM Broadcast Quality

■ Mid-Fi Music Mode
- *4 Tracks/4.8 Hours/Stereo
- *8 Tracks/9.6 Hours/Mono
- *4-Bit Adaptive Delta PCM Encoding
- *FM Broadcast Quality

Fig. 6-6. CD-I audio modes.

And, finally, there is a speech mode, which is equivalent in quality to AM broadcast and is capable of storing 9.6 hours of stereo or 19.2 hours of mono speech.

Text

As for the written word, the CD-I standard provides three storage modes (see Fig. 6-7). The highest-quality mode makes use of bit-map storage and the CLUT compression scheme. This generates attractive text with nice fonts and so forth,

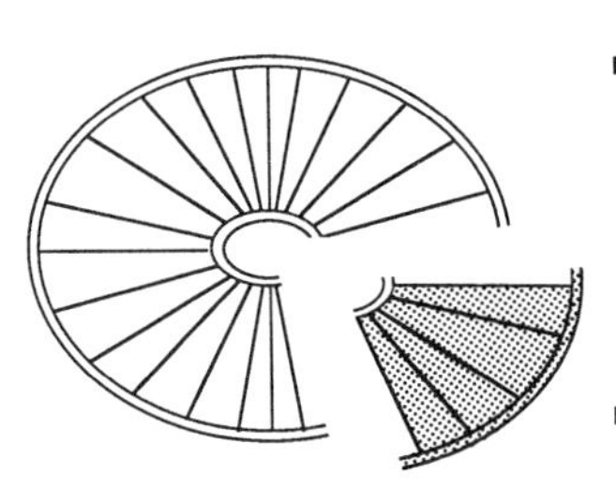

■ Bit-Map Text Storage/Display
- *CLUT Graphics + Compression
- *5 Bytes of Storage per Character
- *120 Million Characters/16 Colors
- *Display Only – No User Manipulation

■ Character-encoded Text (Applications)
- *Character-Specific Attributes
- *2 Bytes of Storage per Character
- *300 Million Characters
- *User Can Cut/Paste and Manipulate Text

■ Character-encoded Text (System)
- *System Standard Text Display
- *1 Byte of Storage per Character
- *600 Million Characters
- *User Can Cut/Paste and Manipulate Text

■ Hardware Digital Storage Decompression

■ Normal TV Presentation
- *40 Characters/20 Lines
- *320 x 210 Pixels

■ HDTV Presentation
- *80 Characters/40 Lines
- *640 x 420 Pixels

Fig. 6-7. CD-I text modes.

but it takes up room (some 5 bytes per character, which means you can store only 120 million characters on a disc) and results in a graphics-based text display that the user cannot manipulate. This storage mode could be used for illustrated books that were meant to be read but not copied.

If you want the user to be able to cut and paste the text you are storing on your CD-I application, consider character encoding. The character encoding scheme has two modes: fancy and plain. The fancy text mode uses 2 bytes to store a character; hence, each character may have attributes such as color, typeface, size, and so on. You can squeeze only 300 million fancy characters on one disc side, however. If you really have a lot of text that you want the user to browse through, use the plain text mode, which will store 1 character per byte and allow more than 600 million characters per disc side.

DESIGNING FOR CD-I

The development of CD-I applications differs from the development of interactive videodisc applications in several fundamental ways, and a designer should be aware of these differences at the outset. For one, CD-I is an entirely digital medium, and, as such, all information is stored as a set of digital data and can be retrieved and manipulated by the microprocessor. Thus, videodisc system designers who come from the analog video world will be faced with new potentials.

On the other hand, because of the low data transfer rate from the CD drive, the CD-I designer is faced with some new constraints requiring new design solutions. A good example is the case of global (screen-wide) dissolves and wipes, in which the beginning and ending images are stored on the disc, along with the instructions to the player for performing the effect. These dissolves and wipes will now be done by the CD-I hardware at playback time, rather than by the editor during postproduction, after which the entire edited sequence must be stored on disc.

Again, it is important to stress that there is a single data stream coming from the CD-I disc, and, in order to properly achieve the capabilities of the medium, the data stream must be carefully structured as a mix of various kinds of information. This goes beyond the "disc geography" concepts developed by the interactive videodisc community, since the data for video, graphics, and several channels of audio that need to be accessed at the same time must be broken apart into 2K blocks and interleaved, rather than just placed in close proximity to each other in complete sections of data.

The unravelling of this data stream, known as a *real-time record,* will be done by CD-RTOS, aided by data-type interrupts buried in the data blocks and the DMA circuitry that will actually deliver the data to the appropriate area at the right time.

Some Potential Applications

The capabilities inherent in the CD-I specification allow for a wide range of applications software to be developed. I will refrain from using the Level I/Level II/Level III classification system until a clearer picture of applications can be developed; however, I will try to touch on the possible types of applications and set forth certain genres.

The near future will bring what I call *ambient video* or *ambient audio* applications, in which a random selection process is applied to an audio database and, in some cases, to a video still-frame database as well. These applications are fairly simple to design and develop, but they offer a novel, nonrepeating, endless listening or watching experience. You can turn them on and leave them on all day. They remain interesting because the user cannot anticipate the order in which the selections will be played.

A similar product, popular in Japan, is the Karoke *sing along*, a concept in which the "performer" sings the words of a song in time to the music provided by a CD or tape. It remains to be seen whether this kind of application will catch on in the United States, although some observers think that it is a natural with country and western music fans.

The *talking book* concept couples the audio and text storage capabilities of CD-I in an application that would teach English and spelling, or a foreign language. This genre would also include "Sesame Street" kinds of applications that would combine linear presentations with interactive practice sessions, and could be stretched to include speed-reading courses.

Painting programs, most of which owe some design allegiance to MacPaint™ on the Macintosh, will also certainly be among the early applications. CD-I versions could be expected to include a lot of audio support and routines to teach techniques; they would also include vast quantities of clip-art to cut, paste, and color. One concept is to digitize Ansel Adams's work, well-known for its use of grey scales in the black and white medium, and then to provide a colorizer palette for each level of the grey scale.

A related product would be a large database of sound effects with a nifty graphic cut-and-paste method of selection and a good indexing scheme—a sort of *audio painting* program.

Couple the current interest in trivia with the large storage capability of CD-I and you can create a whole class of *trivial storage* games. One example might use video and audio capabilities to present an interactive version of a TV game show based on categories of trivia. A random selection process would ensure that no two games were ever alike.

This last product would require only a change of user interface and database to turn it into a study guide for the SAT or one of the other kinds of entrance exams. With a study mode and a test mode, such a program could create an endless supply of sample tests, no two alike. This and other applications that provide the user with access to large amounts of meaningful data might be considered "nontrivial storage" applications.

All the applications mentioned so far use one or two of the CD-I storage modes more or less exclusively and, as such, will be among the first wave of applications created. A second round of applications should begin to balance the use of all media, and the farther you go into the future, the more computer software will be stored per application and the more impressive those applications will become.

A typical midterm application is the fully developed *interactive video adventure game*, with involved plot, multiple characters, and many logical puzzles. The main difference between these games and current microcomputer versions is the fact that they will have made the transition to video from graphics, and the audio will be lifelike rather than computer-like. A variation of these games is the *surrogate travel* experience, in which the user can wander at will in a world composed of video stills. The surrogate travel experience can also be an excellent metaphor for a database if the area being explored is a museum or a collection.

Activity simulation is an application closely related to surrogate travel and the video adventure game. By using design ingenuity, there is very little that an interactive videodisc designer cannot simulate, either for entertainment or for informational purposes, and CD-I is just as versatile. The lack of full-motion video is a limitation in some instances, but certainly not one that cannot be effectively overcome. One example of activity simulation is the flight simulators that are so popular in the microcomputer world.

As mobile CD-I systems become available, first for cars and then for boats and aircraft, there will evolve both *navigation-aid* and *tour-guide* discs. A well-designed tour-guide disc of Bryce Canyon National Park, for example, would play well at home as a surrogate travel experience and then work equally well in the car when the actual tour of the canyon was taking place.

Personal video editing is only as far away as the CD-I systems designed with the capability of controlling VCRs. These systems would allow the home video production to look as polished as professional productions, with switching, windowing, and titling done by the CD-I player as directed by the user via a simple graphics interface.

In the future, as CD-I systems become more powerful, a number of new applications classes will develop. For example, the *video persona user interface* provides a personality user shell, in which a program or other computer system can use a set of standard software calls to communicate with the user. These standard calls will then be translated into a video and audio experience based on that specific disc. If the user gets tired of, say, the stuffy manner of the "Jeeves the Butler" persona disc, then it is merely a quick trip to the CD-I store for a copy of the new "French Maid" disc, and her personality replaces that of Jeeves in the home-control system, car-navigation system, personal robot, or what have you.

The video persona user interface is a good candidate for an "expert systems" front end that uses low- to medium-level artificial-intelligence capabilities to provide users with decision and data-access support. Such an expert system with a video persona could be very amusing at parties as a conversation partner, or it could provide serious, valuable advice to help the user in decision-making.

Another concept that has been around for a while is a *clipping service*. After being directed by the user with regard to the kinds of information desired, the program peruses either data broadcast to the home or external data banks and presents the user with a custom response. The collected information might take the form of a daily "private newspaper," custom-built during the night for morning reading, or it might be presented as a formal report in response to a specific request.

The indispensable *personal computer* of the early 1990s might be a CD-I player with a touch-sensitive flat screen, two small high-performance speakers, enhanced memory, a healthy dose of nonvolatile RAM, a floppy disk, and a modem port. It would run a very software-intensive application combining the best aspects of a video persona, an expert system, and a software agent. Everything would be packaged in a sturdy housing roughly 8½ by 11 by 1½ inches. The back of the housing would be covered with solar cells to charge the player's nickel-cadmium batteries. Such a device could learn about the habits and patterns of its owner as it kept track of schedules and events. It could also simplify and organize myriad details of that individual's life. Not inconceivable is a situation in which two people, relaxing together in a small bar after a hectic day in the city, might agree to connect their micro assistants and run a program in each that would compare habits and patterns and report back to their respective human owners on the potential for mutual compatibility, with estimates for overnight, longer-than a-month, and longer-than-a-year.

All of these applications and thousands more are possible in CD-I. Which ones come out when, and which ones will be commercial successes, are all that remain to be seen.

At first, CD-I efforts will no doubt include audio with simple video scenes or graphics added, since that is a logical first step in learning to produce in the medium; but some thought and innovation could result in a unique application. Either the graphics that support the audio, or the audio itself, could be selected by a random process to provide an endless visual or audio experience. This would make a CD-I record more interesting to listen to than a conventional one. By using the lower audio quality levels, old radio programs of various eras could be stored and then played back in an unpredictable order. The programmability of the CD-DA player could be emulated with a graphics-based user interface, so that program selection and ordering would be done entirely with a point-and-click device.

Initial text-based products might be *databases with indexes* to ensure speedy retrieval, or books translated into electronic text with built-in cross-references and dictionary-type aids. Brown University has done extensive work in this area, particularly in the field of *hypertext*, a method by which a user can build a document by linking together several passages from a large text base.

A CD-I Application for Chess

The real CD-I applications will be those that combine all attributes of the medium into a coherent design that offers many kinds of experiences, depending on how involved the user decides to become with the information. There is a principle that the CD-I application should be ready to respond to a user's curiosity with as much content and depth as can be imagined, but still be entertaining without forcing the user to do anything at all.

To illustrate this principle, let's consider an example of a CD-I consumer design. Consider the case of a CD-I disc devoted to the subject of chess. In conventional media, a product based on chess might take several forms. A book, for example, might review the history of chess, describe great games or players, or focus on certain stages of the game, such as the opening or the endgame. A series of books by the same publisher might reference each other. In the case of an audio cassette, a follow-me-through approach might be taken to discuss how to play the game, with instructions to the listener to move the pieces on a chessboard and stop the cassette player from time to time to consider the options at a given point. A video program, either broadcast or on videotape, might offer an entertaining look at the history of the game, or an analysis of great players and matches. And, of course, a computer chess program could play the part of the opponent or monitor the progress of a game between two human players.

Now let's think about a CD-I disc based on chess. Such a disc would include all of the components previously described and would integrate them completely, so that each was available to support the others. The entire experience would be unlike one that could be had in any other medium. Its approach to its subject matter would be in the form of several modes.

First of all, there would be a *passive* video presentation mode to describe the history of the game, present the rules, and show a sample game in animation. The interesting part of this segment would be that every time you played it, the sample game would be selected from a group, so that you wouldn't see exactly the same presentation twice. (In other kinds of CD-I applications, this passive mode would be an endless, nonrepeating loop.) Although well-structured as a passive linear presentation, the computer program would constantly monitor the input device, and the user could interrupt the presentation at any time and have the option either to study some text or other information, or to perhaps attempt to play out the remainder of the sample game by selecting a side and playing against the computer program that acts as the opponent.

The second mode would be a *browse*, or semi-interactive, mode. In the browse mode, a metaphor for the information contained on the disc would be presented in visual form, and the user could browse through that metaphor in search of a specific area of interest to study. For example, the browse mode might give the user the impression of walking through a chess museum, with the direction and speed of the walk controlled by the user. Any object of interest could be examined to reveal further detailed information, typically dropping the user into one of several chess books stored on the disc.

Next would be a *graze* mode, which would allow deliberate study of information. This could be accomplished by including a few illustrations and a carefully selected set of chess titles as text. These would be cross-referenced, and, whenever a text reference was made to a specific game, player, or event in chess history, the visuals and audio used in the passive mode could be accessed by the user as support for the text. This would be sort of the inverse of the browse mode in a visual/text design sense, as the metaphor here is of reading a book. A sophisticated cross-reference scheme would let the user wander between titles in the set of books and even link together information from them to create a new document for future reference.

Also included would be an *interactive* mode—in this case, a good computer chess program that would have a choice of 3D perspective graphics or conventional 2D chess-notation presentation. The chess program would be capable of playing any of the great games of history or of providing practice on openings or endgames using examples from the great games. A complete formal competition chess match could be simulated, or a tutorial mode could be invoked in which the audio of the teacher's voice would assist the novice with various concepts of the game, or even the special cases of the rules such as *en passant* capture or castling.

Finally, a *highly interactive* mode would be included, featuring an arcade-style game or a 5-minute time limit game. An example of this is the impressive

Through the Looking Glass chess arcade game that Steve Capps wrote for the Apple Macintosh computer.

The structure and linkage between these various modes would be designed to be totally transparent to the user, so that the transition from one mode to another was unnoticed and the interactivity was consistent with the user's whim. In such a disc, the entire gamut of experiences, from passive to interactive, and from informative to entertaining to absorbing, would be offered in a single seamless design.

User Preferences

One convention that all CD-I applications should respect is that of user preferences. A set of parameters stored in nonvolatile RAM could be set by the user to indicate text size and color preferences, and could be referred to by application software. This would ensure that if a user likes to see slightly larger than normal text, he or she would have to explain that fact only once to the operating system and all subsequent applications would see that it happens. User preference could include background color, text color, and colors for text linked to other experiences (such as more text, audio, animation, or software examples).

Standard User Interfaces

Currently, no standard user interface has been defined for CD-I, but it is hoped and expected that this critical portion of the CD-I standard will happen soon. There are several user interfaces written for the 68000 microprocessor, and one of them should be agreed on by the software community as a de facto standard. The gain here is twofold. First and most important, the users of CD-I can develop a consistent set of habits that they can transfer from application to application, speeding the consumer acceptance of an interactive system. And second, application developers will not have to waste precious time developing user interface routines; rather, they can simply call the standard routines to present menu options and engage in dialogues with the users.

CD-I APPLICATION DEVELOPMENT METHODOLOGY

The process of CD-I application development is probably comparable to Level III videodisc system development, although the microprocessor is on board the player, making it in some sense a Level II system. (Level III videodisc systems are those that connect a videodisc drive to a general-purpose micro, in which the control program is resident.) Level III applications are usually much more sophisticated than those found on Level II videodiscs, where the program is resident on the videodisc and is read into the RAM in the player and executed by an on-board microprocessor. A typical Level II player has an 8-bit micro and a limited amount of RAM, and often the program complexity is limited to controlling branching via a keypad. With a 32-bit 68000 microprocessor and a megabyte of RAM in a CD-I player, however, there are virtually no limitations on program complexity or size.

As a result, the CD-I program should be thought of as a computer program that can utilize video still-frames and effects as building blocks to interactivity—along with audio, graphics, and text—rather than as a video program that the user can cause to branch. While much has been said about keeping programmers out of the design process, and about developing tools that will allow the creative artist to design the program, it is this writer's contention that to fully utilize the tools offered by a medium such as CD-I, the program designer should have personally written enough program code to understand both the design potential offered by the system and the amount of effort required by the software production team to achieve design goals. Teaching a general-purpose microcomputer to play a top-quality game of tic-tac-toe using a language such as BASIC is a task that can provide the experienced linear video producer or director with a fair appreciation of the programming process. It is not entirely trivial, nor is it so daunting as to preclude completion by the novice programmer. The process of thinking through how the program should select its next move will conceptually prepare the CD-I designer for using the 68000 to full advantage.

One key concept is to design the program so that image production can be done in still-frames. The designer should also plan on transitions and effects being done in hardware by the player at the time the viewer sees them. The digitized still-frames and calls to the operating system can be thought of as the building blocks of a presentation, and the 68000 can be thought of as the little fellow that will put them together for the viewer as dictated by the circumstances of the control-flow of the program and the user's input. This approach allows the images to be connected in different orders and saves disc space. The designer can simulate the visual flow of the storyboard by careful selection of still-frames and transitions, and will probably be rewarded with a nicely crafted sequence. On the other hand, a video production designer who thinks in terms of 30-frames-per-second video will inevitably be dissatisfied with the result (at least until the advent of the Omni Player).

Another major change in methodology from videodisc design and production is the change to digital data from analog. At some point in the development process, everything that goes onto a CD-I disc will have to be digitized and stored as a set of digital data. The next step will be a transition to the OS-9 file format, and after that the interleaving into a real-time record. From there, the digitized data will be transformed into the CD-I premaster format. The important point to realize is that the transition to digital from analog is inevitable and that there are certain things that you can do with digital data that you cannot do with analog, and vice versa. The sooner the designer understands the differences and the implications of those differences, the better for both the design and the production.

While much talk these days is centered around the need for, and the development of, "authoring systems" that will automate to various degrees both the design and the production processes, the fact of the matter is that, as always, those who complete the first applications will have done so by carefully studying the details of the CD-I specification and by handcrafting a premaster tape. Since the premaster tape has an entirely digital format, it must be built up using a computer to order the digital information, rather than a videotape recorder as in videodisc. As a result, in the early days the development team will have to include a computer systems operator and a proficient programmer just to produce a premaster tape, much less develop the program code.

The remainder of this section will attempt to point the way for those interested in being among the very first out the door with a CD-I application—the enthusiasts who operate on the principle that the early bird does indeed get the worm. For those who would rather wait and see how the market for CD-I applications develops, most of the rest of this section will be obsolete by the time you start developing your application. The good news side of that decision is that much more information will then be available, it will be much more accurate, and both the process and the tools will be far better defined.

The choice of development system is a complex task. It often has more to do with how much existing equipment can be utilized and what expertise is immediately available, than with the demands of the design. Given the very wide variety of potential CD-I applications, a detailed discussion of development systems is beyond the scope of this chapter (not to mention somewhat premature). Several aspects of development system configuration will hold true regardless, however, and will be presented here as a starting point.

First of all, since the CD-I player will be running a 68000-series processor, you will need an assembler or cross assembler for that processor, or you will have to work in a higher-level language that can be compiled at some point into 68000 code. Also, to assemble the final premaster tape, you will eventually need to migrate the source code and data to a system that runs OS-9. (That is not exactly true, as I'm sure that a serious C/UNIX™ guru could produce a properly formatted premaster tape using a VAX or whatever, but unless you happen to have that capability in-house, it's not worth the effort.)

If your development system also runs a 68000, so much the better. However, early conceptual work can be done on an IBM-compatible machine—for example, if the coding is done in the C language or BASIC. Microware offers a fine C compiler that runs under OS-9, as well as an advanced version of BASIC. The C language is easily the language of choice for new concept development work, as it is both the most powerful and the most portable of the higher-level languages. There will be cases in which the existing code is in BASIC, and it is reasonable to leave it there, particularly if you are simply trying to port an existing application over to the CD-I format. You will have enough problems digitizing the images and audio without rewriting the code in a new language.

If you are primarily interested in a graphics-based application and are looking for a low-cost development system, you may want to consider the Atari 1040ST for initial development work. The graphics resolution is reasonably close to that of CD-I, it runs a 68000, and Microware is in the process of porting OS-9 to run on it. Any other kind of application will need an open-architecture computer with a bus, so that the system configuration can be expanded to suit the needs of the design. An IBM clone can be used, and there are cards that plug into it that run a 68000. Hallock Systems Company makes one in particular that also runs OS-9, which would be worth looking into.

If I were starting from scratch, I'd probably choose a Sun III, which runs a 68000-series processor on a VME Bus with a wide-card format, and can be configured with both UNIX System V and BSD4.2 UNIX operating systems. That may sound like a lot of computer, but CD-I application development is a lot of problem to solve, particularly if you intend to store lots of data in your application. My guess is that it won't be too long before OS-9 gets ported to that configuration. But if you are in a hurry and have C and UNIX expertise, you can probably use it to produce a premaster in any event. The Sage/Stride series of workstations are also worthy of consideration.

VideoTools offers a system based on the IBM PC that allows you to develop, simulate, and premaster a CD-ROM disc. That sounds like an obvious solution to the problem, but there is a lot more that you will have to accomplish to develop and premaster a CD-I application, including the 68000 code and interleaving. Nonetheless, for those with strong IBM PC background, that may be the best starting point and may indeed prove to be the fastest path to a development system that could offer some form of simulation for testing prior to premastering. It is also safe to assume that VideoTools will eventually develop its system to include CD-I capability.

In all cases, the limiting factor in the early efforts will be access to information that details the format of a CD-I premaster tape. The very first thing that anyone even slightly interested in developing a CD-I application should do is obtain from American Interactive Media (AIM) a copy of the booklet *A General Introduction to CD-Interactive*, published by Philips New Media Systems. This is the most recent version available to the public regarding the capabilities and requirements of CD-I. Also, AIM is in the process of developing for Philips an application development guide that will probably be available around the end of 1986.

The second thing a prospective developer should do is contact Microware, get information on the current availability of OS-9 on various systems, and buy a copy of Microware's OS-9/68000 Advanced System Software manual. Microware offers a package called a PortPak to qualified potential licensees for evaluation purposes, which includes system object code, I/O driver source code, and all software development tools required to install OS-9 for testing and evaluation on a new hardware system. If you know what you're doing, an OS-9 PortPak will allow you to get your development system up and running in your configuration ahead of the competition. If you don't know what you're doing, it will not be of much use, as it is intended for system hackers, not the application-development types, and you are better off nosing around to see who has already ported OS-9 to your system configuration.

The third thing you should do, assuming that steps one and two still leave you convinced that you want to do your application in CD-I, is to contact AIM and determine the current procedure to obtain a *Green Book*, which contains the actual specifications for CD-I. As of this writing, the release of the *Provisional Green Book* is imminent; but the book will be limited to those intending to provide CD-I hardware, and it is subject to revision. When the final version of the *Green Book* is agreed on (again, not much earlier than January 1987), AIM will be able to tell you how to obtain a copy.

With the preceding information in hand and a development system configuration determined, the process of application design and development can begin. Given the variety of disciplines involved in CD-I, the next step is to assemble a team of experts and agree on syntax and terminology, as well as the application's goal. The basic steps involve (1) design and revision, in which the concept is developed and the interactivity specified; (2) prototyping and production design, in which a model of the application is brought up on a computer system to test interactivity and design concepts and to refine the production design to the point of storyboards and shooting scripts; and (3) production and data gathering, in which the video and audio are assembled in analog form. Following production, the data is converted from analog to digital and assembled in OS-9 file formats, and the source code is migrated to an OS-9 high-level language or 68000 assembly format. Integration and simulation will bring the project to the point of testing and refinement, as the code is compiled and tested in an OS-9 environment. When that stage is satisfactorily completed, development of real-time records, interleaving, and final premastering can take place.

Certain hardware components of CD-I will not be available in first silicon until late 1987, so final testing cannot occur too much before player availability. The lead time remaining until you can actually send your tape off for mastering and place the check discs on your prototype CD-I player may seem more than adequate as of the publication of this book. However, the success of a CD-I application in the consumer marketplace will be won or lost in the design phase, so the sooner you begin to plan the application and test it in some form on a system that can emulate a CD-I system, the better. The most effective aspect of the design process is refinement, which can occur only when you can get your hands on a model of your design and see if it works.

PUBLISHING THE APPLICATION

AIM, based in Santa Monica, California, is a Philips/PolyGram Corporation established to spearhead the development of CD-I application software in the United States and, as such, is the premiere organization in a position to publish CD-I software. AIM will be the first choice of applications developers, since it is directly connected to NV Philips via their parent, PolyGram Corporation, and hence will have the most direct access to the most effective means of developing and publishing consumer-oriented CD-I software. AIM is in a position to set up joint ventures with prospective CD-I software developers and organizations wishing to establish themselves in this field. AIM will tend to focus its efforts on the publication of CD-I consumer software that involves a transition to CD-I from an existing medium; however, that is not a hard and fast guideline. AIM is developing a wide variety of content and applications of the CD-I technology across a broad base of consumer needs.

The Record Group (TRG), based in Burbank, California, is also in business strictly to publish CD-I consumer software. However, the president of TRG, Stan Cornyn, has clearly focused his company's efforts on the entertainment aspects of CD-I. This is understandable, since Cornyn's extensive background in the music publishing business has given him a personal vision of the CD-I player as an unprecedented home-entertainment medium. A number of innovative programs are under development at The Record Group, ranging from primarily audio-based entertainment of the "ambient audio" genre, to sophisticated experiences utilizing all of the power of CD-I in such areas as surrogate travel and entertainment database. While TRG's CD-I publishing efforts will not be as numerous as AIM's, they will tend to involve more creative use of the media. If you decide to approach The Record Group to publish your design, make sure that you have a well-thought-out concept that makes full use of the unique capabilities of CD-I and is geared to the consumer entertainment market. As an innovator in the field, Cornyn's group already has projects under way that many designers are just beginning to consider as possibilities.

KnowledgeSet, based in Monterey, California, is active in the field of CD-ROM and has recently formed Publisher's Data Service Corporation in a joint venture with Digital Audio Disc Corporation, a wholly owned subsidiary of Sony Corporation of America, to offer "one-stop" service for the formatting, mastering, and replication of CD-ROM software. KnowledgeSet specializes in database formatting and retrieval using the Knowledge Retrieval System, and is planning to offer support for CD-I applications. It would be an excellent choice if your application involved extensive data storage and retrieval. The CD-I media, like its cousin CD-ROM, can hold a massive amount of data; however, the access time and the data transfer rate of the drive are generally poor, and anyone planning to take advantage of the storage should take great care to optimize the retrieval process, or the user will not be able to effectively get at the data.

Microsoft has, at this writing, maintained the position that it is interested in developing authoring tools for developers, but such tools will take time to develop and debug. Thus, it is not a choice for the early developer.

The independent applications developer will have considerable problems in creating a CD-I application as compared with, say, a Macintosh application, because of the multiple media nature of CD-I. This is not to say that it cannot be done, but it is not trivial by any means. Even large corporations with considerable resources are entering into joint ventures with organizations such as AIM, in order to have access to the technology and methodology as soon as it is developed.

CD-I is an exciting new technology and one of the most significant developments to date in the ongoing revolution in communications and information management. The challenge to the interactive video systems designer is to create the kinds of applications that will help Philips and Sony realize their dream of a worldwide standard. The reward for those who get it right will be a classic title in a new medium, and the best will eventually be remembered in the same fashion as *Birth of a Nation, The Jazz Singer,* and *The Wizard of Oz.* The marketplace will ultimately determine the fate of CD-I, and the responsibility rests on the shoulders of the initial group of applications developers to make sure that CD-I "sticks to the wall" in the consumer market. If the pioneers of applications development in this new technology are successful, there will be an unprecedented opportunity for those who follow to develop an application for an interactive system with a universal installed base of compatible players.

The choice of whether to lead or follow the CD-I movement is left to the reader of this book, but one thing is fairly certain. If you can't do either of the above, you may as well get out of the way, because CD-I is certainly the Medium of the Future!

7

Controlling CD-I: Languages and Authoring Systems

Marc Canter, Erik Neumann, and Jay Fenton

Marc Canter studied voice, opera, theater, and electronic music at Oberlin College, earning a degree in fine arts in 1980. Further studies at the School of the Art Institute of Chicago included video, kinetic sculpture, and sound. Work in New York City as an audio engineer and in Knoxville, Tennessee, for the Laser Light Show at the World's Fair led him to the video-game market. At Dave Nutting Associates, a subsidiary of Bally/Midway, he worked on Professor PacMan, Ms. Gorf, Ten Pin Deluxe, and Sunken Treasure. He then cofounded MacroMind® with Jay Fenton and has been working ever since to develop state-of-the-art creativity tools.

Erik Neumann graduated from Oberlin College in 1978 with a degree in mathematics and a love of electronic music and computers. While hacking database programs for stockbrokers and oil moguls, he earned his M.B.A. degree at the University of Chicago. Lately, he has been living out the entrepreneurial dream by working for MacroMind. So far he has added MIDI capabilities to MusicWorks™, helped transform VideoWorks™ from an animation program into an authoring system, worked on bringing ComicWorks™ (an object-oriented paint program) to life, and developed the Aquarium™ program.

Jay Fenton started working in the entertainment software business in 1975 for Dave Nutting Associates. There, he helped develop the first microcomputer-controlled pinball and video games. Among his credits are the Bally Arcade home video game system, Bally Basic, and the arcade hit GORF. After the coin-op game market collapsed, Mr. Fenton turned his attention to personal computer software. He worked on Pitstop and Beamrider for Action Graphics, Inc., *and then fell in love with the Macintosh. After cofounding MacroMind, he developed MusicWorks (the first Macintosh music program) and VideoWorks, and helped develop ComicWorks and MazeWars+™. He is currently doing advanced user interface research on the Zorro and Zardoz projects, and is working with Alan Kay on the Vivarium project.*

Now, for those of you who are interested in the "nuts and bolts" of the technology, we present this chapter on CD-I authoring systems.

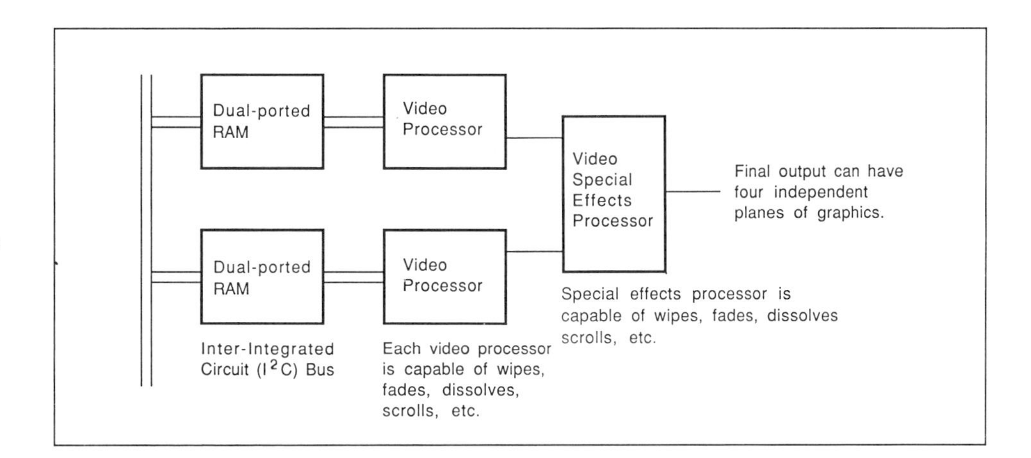

Fig. 7-1. CD-I graphics system configuration.

Since the beginning of time, people have tried to control their environment. And today, since technology is such an integral part of our environment, it is only natural that we should be constantly looking for new and better ways to control the machines we invent to make our lives easier and more enjoyable.

As our machines get more and more complex, developing ways to control them becomes more and more challenging. This is because technology has moved out of the research lab, where it was controlled by scientists armed with programming languages, and into the kitchen, where it must be controllable by the consumer. The average microwave oven or television is a pretty complex piece of machinery, yet the average person can cook a dinner or change the channel with very little training.

Scientists and technocrats didn't choose to hand over control; if it had been left up to them, they would have kept technology locked up in their laboratories. But the sheer momentum of a technology-based economy snatched technology out of their hands and thrust it into the kitchen and living room—and now there is no going back. Thus, while technology grows ever more complex, part of the scientist/technocrat's job is now to make sure that our ways of controlling technology grow ever easier.

CD-I is a technology so complex that we estimate it will probably take at least ten years just to produce one really good application. The specifications for CD-I will be outlined by Philips and Sony in the so-called *Green Book*. The provisional specifications released on June 1, 1986 are completely compatible with the existing CD-DA (PCM) standard. The *Green Book* also has three other audio standards that have been developed to leave room for graphics and data on the disc. These new standards use adaptive delta pulse code modulation (ADPCM) compression so that the audio data is 50%, 25%, and 12% as large as it would be if not compressed.

The CD-I specifications also call for two video coprocessors and an analog signal processor as hardware graphics processors (see Fig. 7-1). The microprocessor is based on the Motorola 68000 family. CD-I also needs a memory management unit (MMU), two timers, two channels of direct memory access (DMA), serial I/O, and 1 megabyte of RAM. All of these features can be found on the 68070.

The operating system for CD-I is called CD-RTOS (compact disc–real-time operating system); it is based on a real-time multitasking system named OS-9. It looks a lot like UNIX and uses primitive drivers to control the CD-I hardware.

All of these specifications are for the *base-case* system. They are the lowest common denominator of features for CD-I. Some applications may take advantage of "extended" features offered by manufacturers choosing to add, say, a synthesizer chip or more RAM. However, all applications must provide at least some sort of default behavior so that their discs can be played on all base-case CD-I hardware. The real issue is how Philips and Sony will build the hooks into CD-I to enable it to work with existing microcomputers manufactured by companies like Apple and IBM, and to work with existing operating systems like Macintosh and MS-DOS.

Controlling CD-I will be somewhat like producing a movie, recording an album, designing an on-line tutorial, writing a book, and programming an application—all at the same time. It is hard enough to imagine a seasoned programmer coordinating all the functions necessary to turn out a product, yet for CD-I to be commercially viable, we have to develop ways of controlling it that will allow people with very little programming background to use this new technology. Can you imagine an inexperienced user controlling a million dollars' worth of hardware? Then you have just imagined what CD-I will be like two years from now.

PROGRAMMING LANGUAGES: INAPPROPRIATE FOR THE TASK

Why, you might ask, can't we just control CD-I the way we have always controlled computers—with programming languages? When you get right down to it, programming languages are simply too hard for the average person to use. Take, for example, the following piece of code, which is written in PASCAL:

```
PNAME:='ELVIS FACE';
DISPLAY(PNAME);
FOR I:=1 TO 200 DO;
VNAME:='HOUND DOG';
DISPLAY(VNAME);
DIALOG('PRISCILLA', 10, 20, 40, 80, 'GRACELAND', 30, 20, 60, 80)
CHOICE:=USERCLICK(MOUSEPOSITION);
IF CHOICE = 1 THEN
    BEGIN
    VNAME:='PRISCILLA';
    DISPLAY(VNAME);
    MNAME:='JAILHOUSE ROCK';
    PLAY(MNAME);
    END
ELSE IF CHOICE = 2 THEN
    BEGIN
    PNAME:='GRACELAND';
```

A nice, tidy piece of code, says the programmer; gobbledegook, says the nonprogrammer.

One problem with programming languages is that society views them, rightly or wrongly, as the scepters of the priesthood of computer programmers. Another problem lies in the attitude of programmers themselves.

Historically, there was some resistance in the computer science community to making the control of computers available to "the masses." When the concept of interactivity first surfaced, it was only begrudgingly given a place next to the printers and punch cards, not because early computer scientists weren't interested in people but because they were more interested in machines and in what people could do with them.

As computer hardware and software improved, however, interactivity became possible in real time (or apparent real time), and it did not take long for computer buffs to realize that interacting with a computer could be fun, that computer graphics and sound effects were powerful weapons in the fight against boredom, and that consumers (particularly young consumers) would pay good money for interactive programs (particularly video games). Suddenly, interactivity became a viable commodity, and the idea that it was important to be kind to the user began to win converts.

The "be kind to the user" type of programming became known as *user-interface design*. In recent years, it has gone from being one of the most ignored and misunderstood facets of the computer industry to spawning a whole new science called human factors research. As a result, most programmers now have to work on their program's interactivity almost as much as (or more than) they work on its internal structure.

Programmers are constantly faced with the problem of coming up with "better" products boasting more features, speed, and versatility. It is not so difficult to come up with products that are more advanced or different, but the trick is to come up with something that is at the same time easier to use. And to do that, programmers have to be willing to rethink a lot of the basic concepts they were taught in computer classes, the most basic of which is that input means typing commands at a keyboard.

There are numerous programming languages available today, most of which still use text strings to represent the functions, procedures, variables, and data that make up programs. This is not to say that programming languages haven't come a long way since the days when programs were encoded on stacks of cards and kept in order by stout rubber bands. The invention of the bit-map display means that we can now see on our screens the programs we create, and developments like Smalltalk, windowing systems, music editors, and the Macintosh have brought computer programming languages a long way. And "languages" like Visicalc, dBASE, and WordStar® (or their descendants) have all become much more commonly used than even PASCAL or C. If we think about such products as VideoWorks, Pinball Construction Set, and ChipWits, we start to realize that the differences between entertainment products and programming languages are crumbling fast.

However, until languages become as easy to use as a microwave oven or a television, computers (and particularly complex computer technologies like CD-I) will remain the temples of the priesthood. Unless, that is, we can develop other ways for people to control them.

AUTHORING LANGUAGES: A STEP IN THE RIGHT DIRECTION

In theory, authoring languages should be easier to use than more conventional programming languages, since they are intended for nonprogrammers, such as writers, designers, artists, musicians, and administrators—anyone who works with ideas to produce multimedia learning tools, entertainment products, or artistic pieces. Authoring languages accomplish this in part by using already familiar terminology. For example, an authoring language for creating slide shows would include commands like *fade, dissolve, switch,* and so on. The Tutor authoring language on the PLATO® system takes this approach in designing courseware, with commands like *course, tutorial, lesson, question,* and *answer.*

In practice, however, authoring languages tend to be nearly as difficult to use as programming languages. This is because they are usually just rehashed versions of the same text-based concept. Also, the nature of the problem—controlling interactive sequences of events through time—is a very complex activity in itself. For example, here is how a simplified "cue sheet" approach to an authoring language might look (this example serves the same function as the PASCAL example given previously):

CUE#	Name	Time	Frame	Link	TimeOut	Button1	Button2
1	Elvis Face	:30	3	2			
2	Hound Dog	1:15	48	3			
3	Which?	:00	90		:30	4	5
4	Priscilla	:15	93	5			
5	Graceland	:45	156	6			
6	Jailhouse	1:30	231	7			
7	Quit?	:00	289		:45	1	8
8	Goodbye	:20	378				

While this approach, which uses lists of cues and pointers to link to other cues, works for videodiscs or slide shows, the situation will become still more complicated with CD-I, where graphics, text, computer code, and sound are all separate entities that must be coordinated very precisely. New approaches to making this kind of information easier to work with will have to be found to make CD-I production affordable. It is much easier for nonprogrammers to work in a "what you see is what you get" (WYSIWYG) kind of world, than one based on symbols that must be memorized and numbers that must be interpreted. This is why the new graphic-, menu-, and icon-based approaches to user interfaces are so important. We will give some examples of graphic-based authoring languages later in this chapter.

CD-I AUTHORING SYSTEMS: THE THEORY

An authoring *language* is a means of specifying the order of events in time and of designing the interaction with the user. An authoring *system* is the total process (including software and hardware) of producing an interactive multimedia experience. In this section, we will talk about an ideal authoring system for producing CD-I applications.

The most important qualities we are looking for in an authoring system are speed and flexibility. Designing an interactive script is very difficult if we cannot quickly see what we have just designed. Authors must be able to see immediately what their ideas look (or feel) like, so that they can alter them while they are still "hot." Any system that requires a lengthy "compile" phase to look at the sequences and try out the interaction will be a tremendous drag on the process.

There are already some good models of this ideal environment for producing CD-I applications. For example, the Edit Droid by LucasFilm is an editing system for producing feature films and commercials. After converting the film to video (only for purposes of editing) and putting the video onto massive computer hard disks, the system lets the editor very quickly arrange sequences of film. It does this by having all the source documents (the videotapes of the film) on very fast hard disks, so that the editor can specify any sort of cut and immediately see the result. The computer system that the editor works with does not actually rearrange any of the source material; it just shows it in a different order. When the editor is satisfied, the computer can then produce a list of edits on paper to be sent to the cutting room where the actual film is cut.

The (as of now) theoretical CD-I authoring system we have in mind is divided into three main sections (see Fig. 7-2):

1. *Domain Editors*—Editing programs that work in specific domains; that is, text, graphics, sound, and computer code. We will use the domain editors to prepare the *source documents* from which we will build our CD-I application.
2. *Authoring Language*—A program that will tie together all the source documents and let us preview the CD-I application.
3. *Emulator*—A hardware device used to test and assemble the final product before production and duplication are begun.

Domain Editors

A variety of domain editors will be available on a number of different computers and operating systems. Before the source documents are sent to the authoring language, most of them will need to go through a conversion step into a standard CD-I format. For example, graphics must be put into either DYUV, CLUT (color lookup tables), or RGB (red-green-blue) format; sounds must be converted to ADPCM, compressed at one of three sampling ratios; computer code must be compiled into the CD-RTOS run-time object code; and text must be put into one of the two standard CD-I formats—text strings or bit-mapped.

The workstations running these domain editors will be separate computers (of whatever brand is best suited for the task: IBM PC, Macintosh, or Sun), networked together for quick trading back and forth of source documents. They could also be remote workstations at home or in another city. The workstation approach is important because it takes a lot of source documents to fill up a 600-megabyte CD-I disc.

Authoring Language

The authoring language is the means by which all the source documents will be tied together into a CD-I application. Thus, the authoring language will have

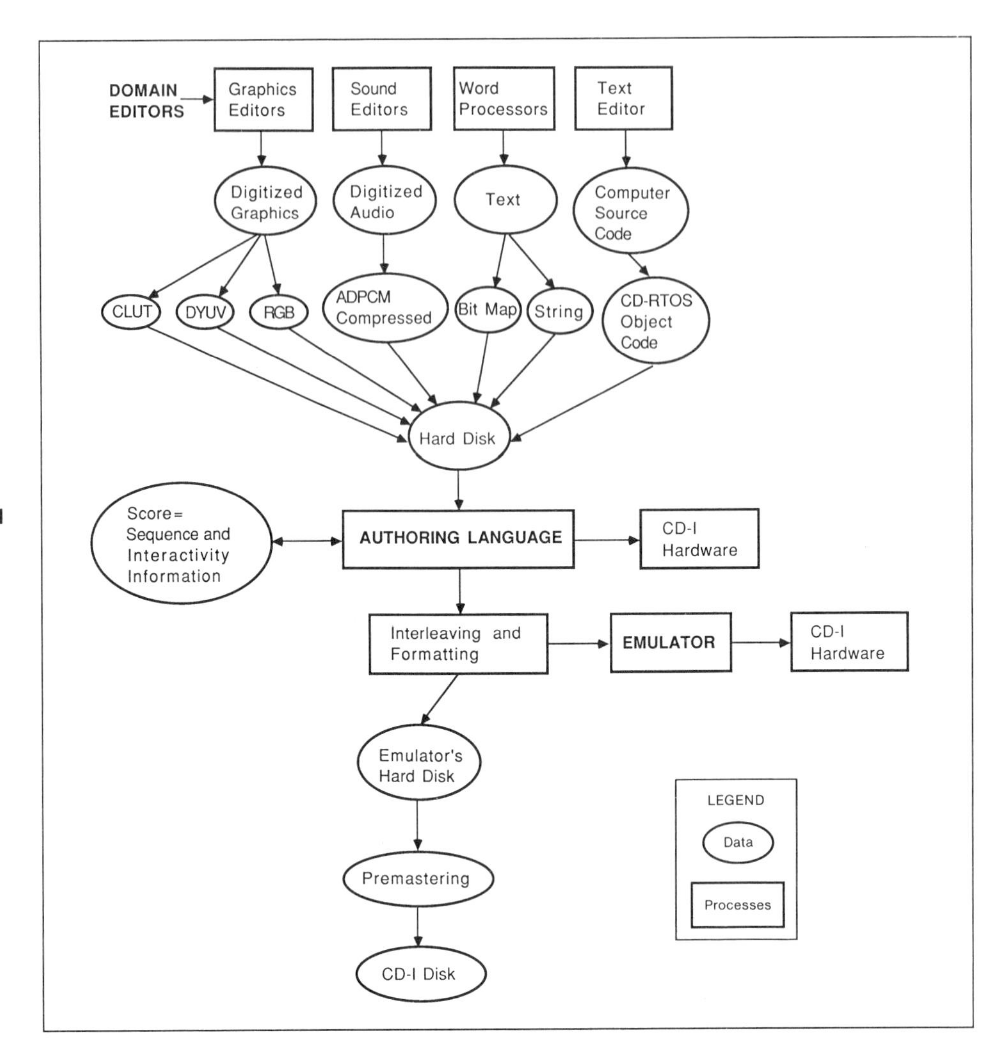

Fig. 7-2. The ideal CD-I authoring system.

to be an advanced, graphics-based language designed to sequence events and to design interactivity. The source documents will be contained on a fairly large hard disk attached to whatever machine the authoring language is on. The authoring language will create its own documents (*scores*) containing the sequencing and interactivity information, and it will provide immediate feedback for the designer by avoiding lengthy compilation as much as possible. It will require a CD-I add-on card to display the CD-I graphics and reproduce the sound, though it would also be possible to substitute standard computer displays.

The authoring language will be written in a machine-independent fashion so that it can be run on IBM PC, Macintosh, or Sun machines and can work with any available CD-I emulators and disc formatters. Imagine an environment in which the hi-res graphics digitizer is on a Sun/3, the MIDI setup is based around a Macintosh, and all text processing and input is done on IBM PCs. Because we will need to be able to pull together the raw data from all the development environments, one way to think about the authoring language is as a database management system (DBMS) that will keep track of the various source files. Our DBMS should be capable of interfacing with all of the development environments.

Besides being a DBMS for source files, the authoring language must also be a DBMS for the CD-RTOS real-time file structure. This means it must interface with the master CD directory that is used to create the CD image itself and work with the lower-level premastering, formatting, and emulation utilities also necessary to produce the CD disc. This master directory eventually gets turned into the directory that is burned into the disc. Each manufacturer who supplies complete development environments for CD-I will have its own type of master CD directory. Currently, Microware, VideoTools, Optical Media, MicroTrends, and a host of others are expected to become CD-I development-system manufacturers, and the authoring language will have to interface with them all.

It will be important for the authoring language to impose some constraints on the application, since, with the fast access times of a hard disk, we could do things with that program that would be impossible with CD-I. The authoring language will also need several sophisticated tools specific to preparing documents for CD-I. Disc layout, preloading, resource management, and interleaving are the sorts of activities that will need to be done.

Emulator

Because of hard-disk space limitations, the CD-I application will probably be prepared in small segments, rather than in its entirety. Each of these segments must then be tested on the emulator, a specialized (and expensive) piece of hardware designed to simulate a CD-I machine. We will use the emulator to try out the complete CD-I application (all the small segments together) and to assemble the final CD-I disc prior to publication.

Advantages of This Ideal System

The advantages of the authoring system we have outlined include the following:

1. Several people can be working on different sections of the application on independent workstations.
2. Whatever domain editor is best for the job can be used on whatever hardware is available.
3. Changes can be easily made to a source document with a domain editor and then can be sent back to the authoring language.
4. The authoring language will be very fast, giving the designer the shortest possible feedback loop.
5. The designer can plan disc layout and interleaving with the authoring language, which will make later steps in the production process easier.
6. An expensive emulator is not needed to develop the CD-I application, only to perform final testing and for production.

Interleaving

Since CD-I applications work in real time, we need to pay special attention to the integration of audio, graphics, text, and computer code. CD-RTOS's real-time file structure uses a process called *interleaving* to make sure that audio and video are loaded and played simultaneously on the CD-I hardware. The authoring language must be capable of organizing our data into this interleaved file structure, and it should help us plan the preloading of audio and video.

Interleaving will convert the separate source files, according to the scheme specified in the authoring language's score, into one long stream of data that can then be used to emulate the CD-I application. Since CD-RTOS can handle information in a noninterleaved fashion, it will be possible to send the graphics, sound, text, and computer code to the CD-I card without having to specify interleaving first. This means that we can put off interleaving during the initial stages, which will save time when the project is still changing a great deal. Eventually, however, we will want to specify the interleaving so that we can test a section of the application on the emulator to ensure that everything is working as it should and so that we can ultimately produce the finished application.

The CD-RTOS file system is specifically designed to slave off of the audio information, so that a constant stream of audio data can be read from the CD-I disc and fed to the audio hardware. In the case of the Type A audio (which is the same quality as CD audio, but half the size), every other block on the CD-I disc will be audio information, with the other blocks left for graphics or other data. In the case of Type B audio (which is half the size of Type A), every fourth block will consist of audio information, with the other sectors left for graphics or data.

CD-RTOS's interleaved real-time file system uses 2K blocks of data, with a sync block on the disc to control the routing of each type of data to the correct hardware. For example, when an audio block comes along, a switcher will physically route the information to the audio hardware, and when a graphics block comes along, the switcher will route the information to the graphics hardware. This type of interleaving enables graphics and data to be preloaded into memory and triggered when the appropriate audio cue is reached on the disc.

Fig. 7-3 is a diagram of an example that could be part of a video game or a simple animation/music combination. In the first section, a large amount of graphics background data is loaded into memory. The background is then displayed at the start of the second section, while a steady sound stream is played. At the same time, foreground graphics for animation is loaded into memory. It is interleaved with the sound stream currently playing. In the third section the foreground animation starts, along with the associated synchronized sound.

Another example is shown in Fig. 7-4. Here, program data is loaded into memory, followed by the graphics for a menu. When the menu is displayed on the screen, it makes four different channels of audio available for the user to choose from.

These examples are generic enough to show you that CD-RTOS's multiple streams of audio, video, and data are a powerful data structure. They also point out the need for an effective way to represent this level of complexity in the authoring language. Because of the extreme time constraints of CD-I's search

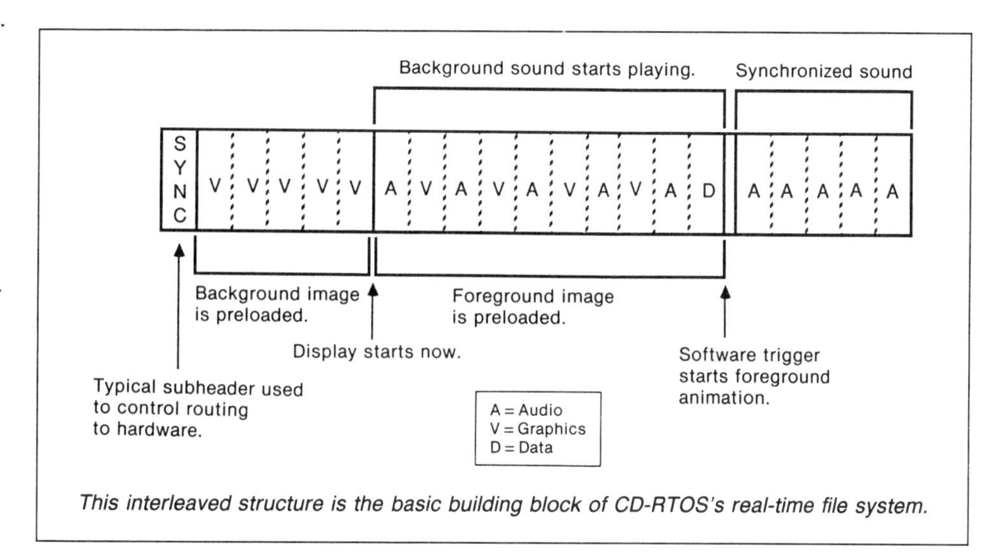

This interleaved structure is the basic building block of CD-RTOS's real-time file system.

Fig. 7-3. A typical sequence of interleaved blocks.

times, the authoring language should make the details of the CD-I interleaved file structure obvious to designers. A circular overview of the disc (something like DNA maps), similar to the one shown in Fig. 7-5, can be used to display the structure of the disc at its highest level. Designers can then zoom in to reveal segments in more appropriate ways, such as ASCII, machine instructions, hexadecimal constants, program instructions, pictures, scores, and so on.

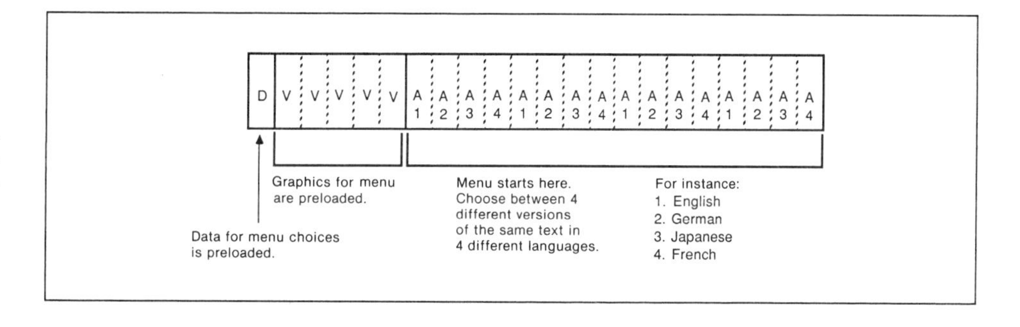

Fig. 7-4. Another example of interleaving, this time with a four-channel audio menu.

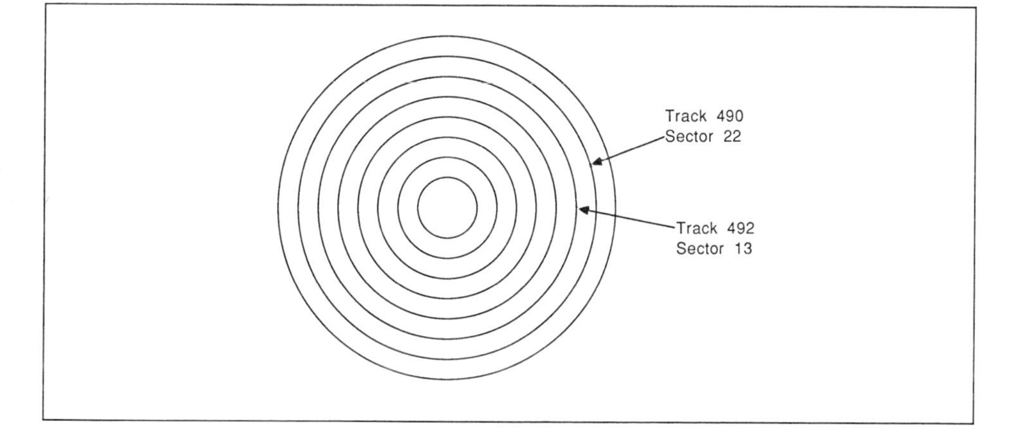

Fig. 7-5. An overview of a CD-I disc, displaying the file structure and disc layout.

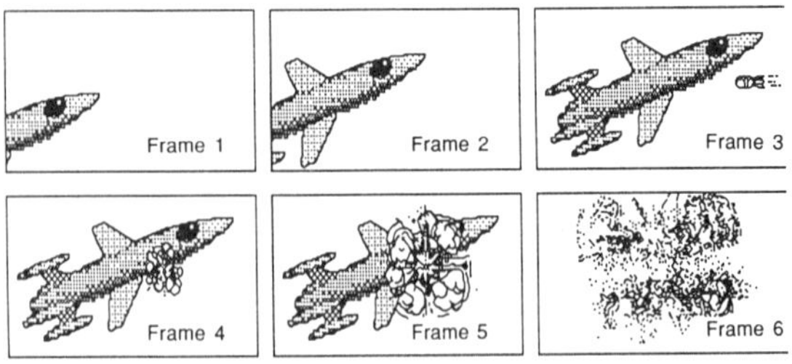

Fig. 7-6. A frame-by-frame sequence from VideoWorks™.

On the other hand, it would also be very good for the authoring system to take a first pass at interleaving and preloading on its own, shielding the author from these details if desired. This would take the form of an automatic "resource manager" that would, for example, ensure that the graphics were already loaded when the audio started playing. The flexibility of being able to switch between the automatic and manual approaches to laying out CD-I on the disc would be very helpful.

CD-I AUTHORING SYSTEMS: AN EXAMPLE

In this section, we will describe an existing authoring system, VideoWorks™, that can be easily adapted to CD-I. It has some of the characteristics of the ideal CD-I authoring system just described: It can take graphics, text, music, and computer-code segments from several sources; it has an animation score that is represented graphically and so is easier to use than traditional cue sheets; it provides immediate real-time feedback to the designer and is fast and easy to use; and it incorporates a complete authoring language for specifying interactivity.

VideoWorks was released in April 1985 as a consumer animation creation tool for the Macintosh. Since then, it has gone through a metamorphosis and is now a full-fledged authoring system for the Macintosh, capable of integrating text, graphics, animation, speech, and music in one interactive application.

Let's first take a look at how VideoWorks handles a simple animation: a spaceship exploding. For this example, we will have the spaceship slide from left to right; then, when a missile appears, the spaceship will explode. Frame by frame, the sequence will look as shown in Fig. 7-6.

To do this sequence with VideoWorks, we first create the images (or copy them from existing art disks) and put them in a collection of images called the *cast*. We need five separate images for this animation: the spaceship, the missile, and the small, medium, and large explosions. Fig. 7-7 shows how these elements look in the VideoWorks cast window.

VideoWorks uses the analogy of an *animation stand* for creating animation. Animation stands are used by Disney and others to do cell animation (frame-per-frame animation). The entire screen is the animating area, and animations

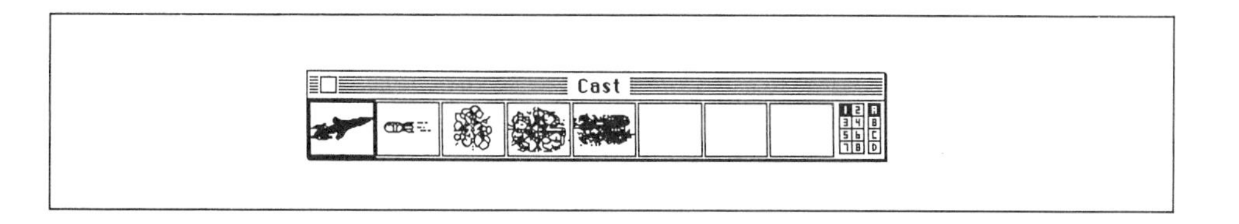

Fig. 7-7. The VideoWorks cast window with five cast members.

can be created by recording the movement of the mouse in real time, or by traditional cell animation (one frame at a time).

To create a sequence, we drag each piece of artwork out of the cast window as we need it, place it on the stage (the screen), and "shoot" each frame of the movie one by one. For example, we first drag out the spaceship and place it in the lower left corner of the stage; then we shoot frame 1. Next, we move the spaceship up and to the left and shoot frame 2. Continuing on, we shoot a total of six frames. The movie can then be played back immediately.

So far this all seems fairly straightforward—not terribly revolutionary at all. But there are two important points to be considered.

First, even after shooting these six frames, the images still exist as independent objects; that is, the frames of the movie don't become fixed-in-concrete images. Rather, they are stored as "position-time" data for the independent objects. This means it is easy to go back and slightly change the way any of the individual frames look, without having to start from scratch.

Second, a powerful notation and editing system, called the VideoWorks score, allows us to look at, understand, and change all of the position-time data. For our spaceship movie, the score looks as shown in Fig. 7-8.

Let's take a moment to understand how the score represents the spaceship movie. What we see in the score is time moving from left to right, represented as frames of the movie (six frames are shown here). The black square in the lower part of the score is called the *playback head*: Whatever frame it is on is the one being shown at the moment.

Although VideoWorks can store up to 512 separate images in the cast window, only 24 of those 512 images can appear on the screen at the same moment. We call each of the 24 images on the screen a *sprite*, and we show them in the

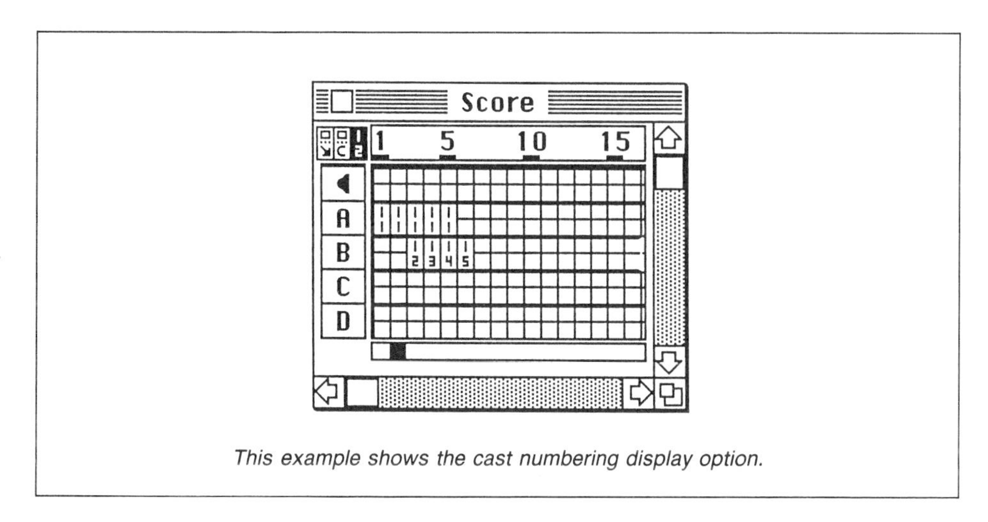

This example shows the cast numbering display option.

Fig. 7-8. The VideoWorks score with two channels of animation.

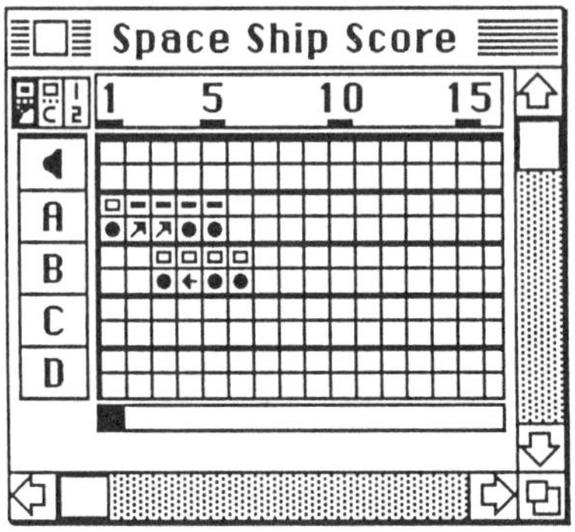

Fig. 7-9. The VideoWorks score with three channels of animation.

This example shows the relative motion display option.

score as one of the stacked rows A through X (the channel above A is the sound channel, which is used to add sound effects to the animation). Therefore, in this movie we are using a maximum of two sprites at any one time; that is, the spaceship in channel A and the missile and explosions in channel B.

There is one more piece of information being shown in the score: The numbers in the little boxes specify which *cast member* (that is, which image) is being shown in this sprite for this particular frame. As you can see, sprite A is always the first cast member—number 1-1, which is the spaceship. Sprite B changes for each frame, starting with the second cast member (number 1-2, the missile), then the third cast member (number 1-3, the small explosion), and so on with the other explosion images (numbers 1-4 and 1-5).

Now you can appreciate how the score graphically represents not only each frame of the movie but also the separate images composing that frame. Of course, all the standard Macintosh editing commands work on the score. You can select pieces of animation, and then you can copy and paste them anywhere else in the score. Many other advanced editing features are also built into VideoWorks that make the life of the animator much easier.

There is still more information available from the score: By clicking one of the *score display switches* in the upper left corner, we can see the relative motion vectors of the sprites from frame to frame, as shown in Fig. 7-9.

The spaceship (sprite A) is moving up and to the right, so the relative motion vectors point in that direction for frames 2 and 3. The arrow shows how the sprite has moved since its position in the previous frame. The dot or circle shows that the sprite has not moved at all, which we can see for frames 4 and 5 of the ship. Other information that can be gleaned from the score includes graphics special effects (write modes) and action codes.

Adding Interactivity with Action Codes

Now that we have seen how easy it is to use animation with VideoWorks, we can introduce the concept of *action codes*. Bear in mind that we are no longer talking about the consumer version of VideoWorks, but a professional-level

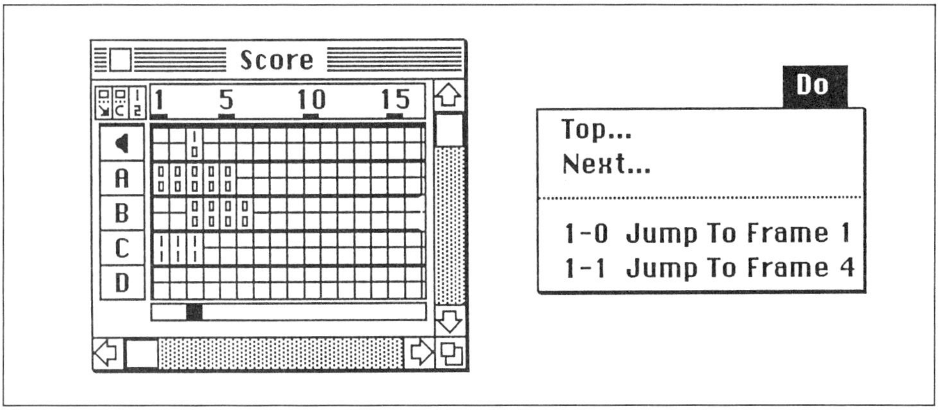

Fig. 7-10. The VideoWorks Do menu and the score with three channels of animation and an action code in the sound channel.

product currently being used by several major software and hardware companies to create guided tours of their products and training courseware. This application shows the potential of VideoWorks for use as a CD-I authoring system.

There are many different types of actions available in the VideoWorks Guided Tour Authoring System. The simplest type of action is to click a button and jump to another frame. For example, we might give the user the options "yes" and "no" as answers to a question. Where the tutorial sequence goes next depends on the user's response, and is implemented by jumping to a different frame for each answer.

In VideoWorks, actions are added to the score by selecting the sprite that they will affect and then selecting the action from a Do menu.

Fig. 7-10 shows the same score as the one shown earlier, but now it is displaying action codes. There are two classes of actions in VideoWorks: those that happen immediately (as soon as the animation reaches that frame) and those that happen when you click on a sprite. The immediate actions are shown in the sound channel, while the others are shown in whatever sprite channel they are associated with.

The 1-0 action code means "jump to frame 1" (the 0-0 action means "do nothing"). Since the jump action appears in the sound channel, it will happen immediately. So what we have here is a loop. When the movie is started, it will play frames 1 and 2, and then the action in frame 3 will take effect and we will jump back to frame 1. We will see the spaceship approaching over and over, in a sort of time warp.

To get out of the time warp, we have added the button shown in channel C of the score in Fig. 7-10, which will look as shown in Fig. 7-11. The 1-1 action code in channel C and in the Do menu means "jump to frame 4." When the user clicks the button, the sequence will immediately jump to frame 4 and the explosion will take place.

To actually do all this in VideoWorks takes a lot less time than you have just spent reading about it. You can even enter the action codes while watching the animation.

You can now imagine how these simple building blocks can be used to quickly build up complex interactive instructional documents. Many other action codes besides simple jumps are available. Some examples are: text boxes that the user can type into; if/then tests of the user's actions; subroutine jumps;

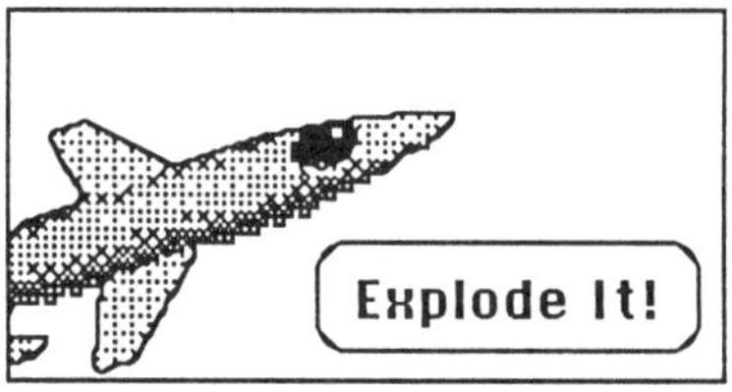

Fig. 7-11. Frame 2 with an interactive button.

Pushing the Explode It button causes the animation to jump to frame 4 and start the explosion animation.

dragable objects; custom Macintosh-style menus (that start any action code); launching outside applications; timeouts for when the user goes to sleep; and many more.

The Macintosh is one of today's most important proving grounds for new approaches to user interfaces. The VideoWorks approach to designing user interaction and integrating text, graphics, and sound—which was developed on the Macintosh—will be important for designers and producers of CD-I software, since immediate feedback and ease of editing will drastically reduce the costs of producing CD-I.

THE FUTURE: ZORRO AND ZARDOZ

Some of the future possibilities of CD-I are being realized in two new products from MacroMind: Zorro and Zardoz. Zorro is the latest step in the evolution of the VideoWorks user interface. It is a sort of flowcharting overview tool. Eventually, Zorro and VideoWorks (along with MusicWorks) will merge, creating Zardoz, an integrated multimedia authoring system.

Zorro

To get an idea of Zorro's usefulness, imagine that you are in the beginning stages of a CD-I project. You have a number of source documents in the form of graphics, text, and animation sketches. At this point, you want to quickly pull all these materials together and add some user interaction so that you can get a feel for what your final product will be like.

Zorro is perfect for this kind of experimentation. Instead of requiring you to type in lists of numbers that represent pointers and jumps to different files, Zorro provides a graphics environment in which you can flowchart your application in a very natural way.

Zorro will display MacPaint™ or MacDraw™ graphics documents, VideoWorks animations, and MusicWorks songs (MusicWorks is MacroMind's music program). Eventually it will also display other sorts of documents, such as MacWrite™ text documents. To specify what documents you want displayed, you drag out the appropriate icon from the Zorro icon window to the document window. Each icon is associated with one of the particular display programs (for example, MacPaint, MacDraw, and VideoWorks). You then type the document's name under the icon and arrange the icons in whatever order you want.

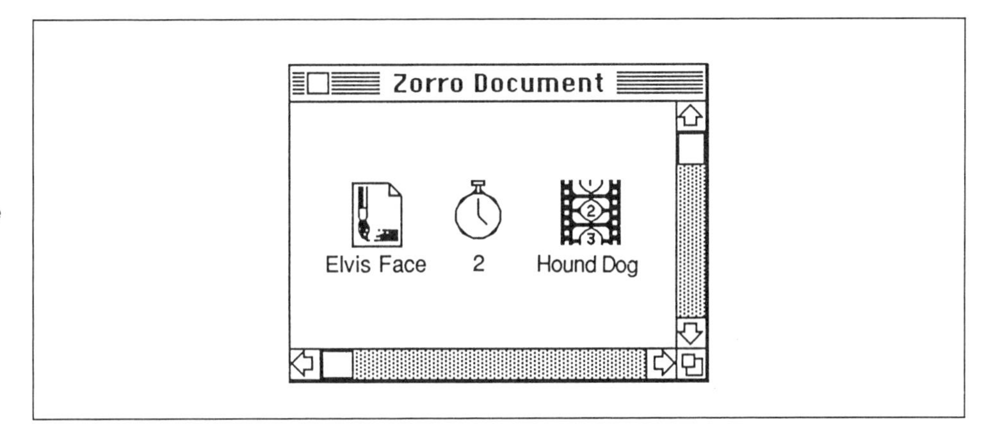

Fig. 7-12. A simple Zorro sequence.

Zorro also has a timer icon, which looks like a stopwatch, for setting a time delay between graphics documents. You type the desired amount of delay, in seconds, under the timer icon.

Let's take a look at an example to see how easy it is to model an interactive environment. Time proceeds from left to right in Zorro. Fig. 7-12 shows a sequence that will first display the Elvis Face MacPaint picture, wait for 2 seconds, then run the Hound Dog VideoWorks animation, and stop.

But suppose we want to add some interaction. Let's say we want to ask the user a question by placing two buttons on the screen after the Hound Dog animation. Depending on the response (which button was pushed), we then want to take one of two different paths. This modification is shown in Fig. 7-13. For clarity, we show the icons connected together with lines. Connecting the icons is as easy as clicking and dragging the mouse between the two points you want to connect.

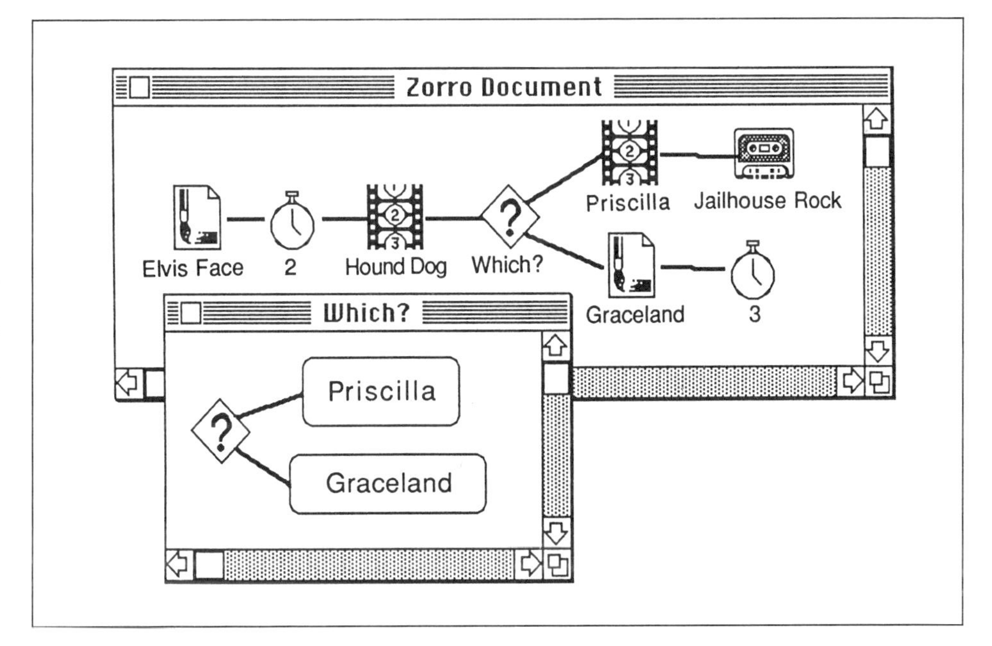

Fig. 7-13. An interactive Zorro sequence with the *Which?* question window displayed.

Fig. 7-14. A sequence with two question icons.

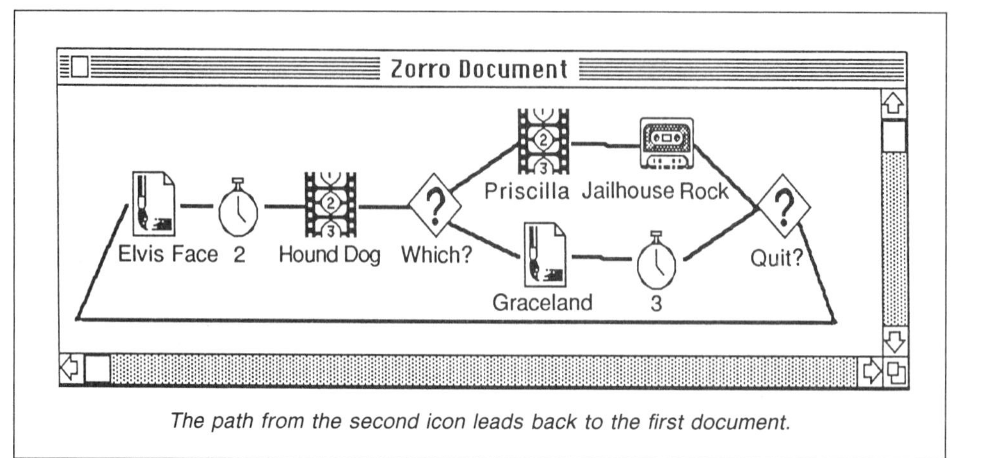

The path from the second icon leads back to the first document.

As you can see, we have added a question icon named *Which?* to this sequence. Double-clicking the question icon will put on the screen a question window displaying two buttons, one with the name Priscilla in it, the other with the name Graceland. Each button is associated with a particular path leading from the question icon. Depending on which button is clicked, our example will either show the Priscilla animation and play Jailhouse Rock or show a picture of the Graceland mansion for 3 seconds.

Now suppose we want a final question, like the one in Fig. 7-14, that asks if the user wants to loop back to the beginning of the sequence. We can end the two separate paths with a *Quit?* question icon and draw a line leading back to the beginning from the question icon to create the loop.

Using these basic elements, very complex sequences can be created rapidly. To keep the complexity from getting out of hand, you can create "folders" within Zorro to hold a sequence. Lines lead in and out of the folders, but as shown in Fig. 7-15, the details of their sequences are not shown until you need to see them.

Fig. 7-15 shows a fairly complex interactive sequence which, again, could be created a lot faster than it takes you to read about it. This time, three paths lead from the *Quit?* question icon: One leads to a folder named Ed Sullivan, another closes the program and returns the user to the Finder, and the third loops back to the beginning of the sequence. You can see in Fig. 7-15 that the Ed Sullivan folder contains a love scene consisting of a Sunset animation and then the song "Love Me Tender." Once the folder has completed its sequence, it returns to the *Quit?* question icon.

Another nice feature of Zorro is that, using it in combination with the Switcher (written by Andy Hertzfeld) on the Macintosh, you can go directly from Zorro to any of the applications used to create or modify your source documents (for example, MacPaint, VideoWorks, or MacWrite). Double-clicking an icon launches that particular application. This creates a superintegrated environment for creating interactive documents.

Zardoz

Formerly called SoundVision, Zardoz is more than a CD-I authoring system; rather it is an integrated multimedia authoring system that can handle 600

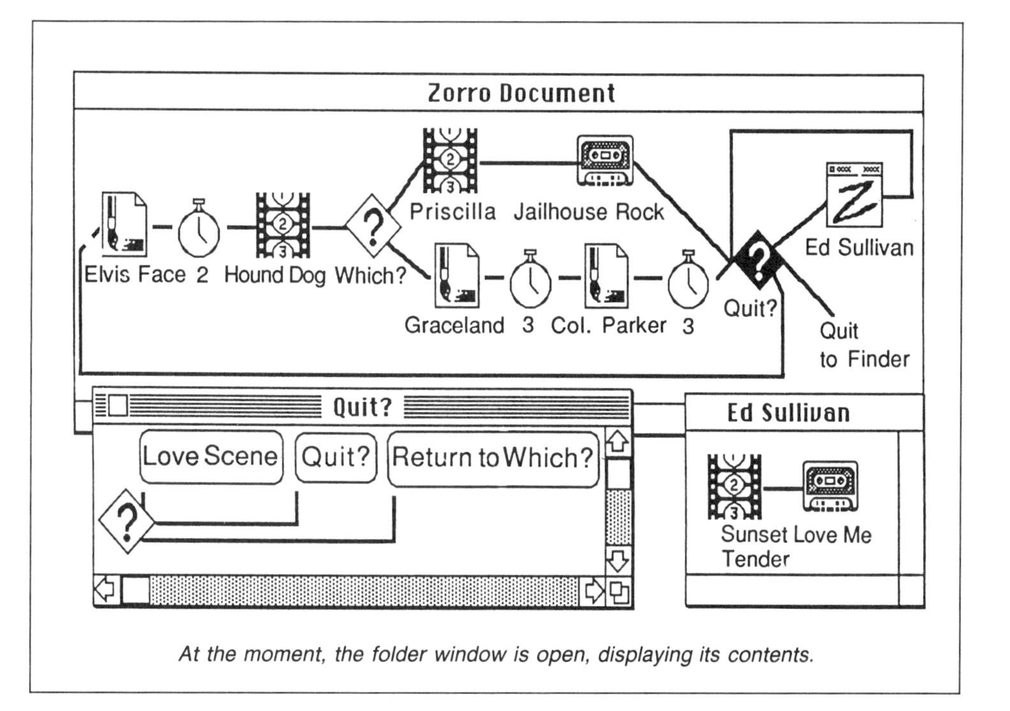

At the moment, the folder window is open, displaying its contents.

Fig. 7-15. One of the paths from the *Quit?* question icon leads to the Ed Sullivan folder.

megabytes of information as easily as 0.6 megabyte. Zardoz will be a courseware generator, a front end to professional video and audio systems, and a front end to Alan Kay's Vivarium project (see *MacWorld*, August 1986, for more information on Vivarium).

Generally speaking, Zardoz will be phrase and scene oriented. It will feature various viewpoints on the data that will let designers work on whatever level they need to; for example, the envelope of the sound, the path of a flying bird, or the order of an entire animation sequence. The system will consist of a wealth of domain editors, each of which will handle a particular type of data, be it graphics, sound, or text.

A "meta-editor" will act as a master notational score, which on at least one level will look a lot like the current VideoWorks score. However, the score will be expanded to include a comment field for typing in action codes directly or for labeling frame numbers, multiple actions per frame (paragraphs of code with While and Case statements), and more detailed information on the animation, as shown in Fig. 7-16.

Zardoz will notate music and analog voltages just as easily as VideoWorks notates animation right now. Synchronization between various media will be accomplished with a graphics notational language named Hookup that features icons that are wired together, similarly to Zorro. Hookup, which is being developed at MIT by David Levitt, Mike Travers, Bosco So, Ivan Cavero, and Bert Sloane, will complement Zardoz's notational approach to programming with an algorithmic approach. Instead of notating events through time by rote—in other words, begin with this, have a middle with this, and end with that—Hookup will have a program (or circuit) that will, for example, constantly change the setting of the color registers or create a sound drone in the background.

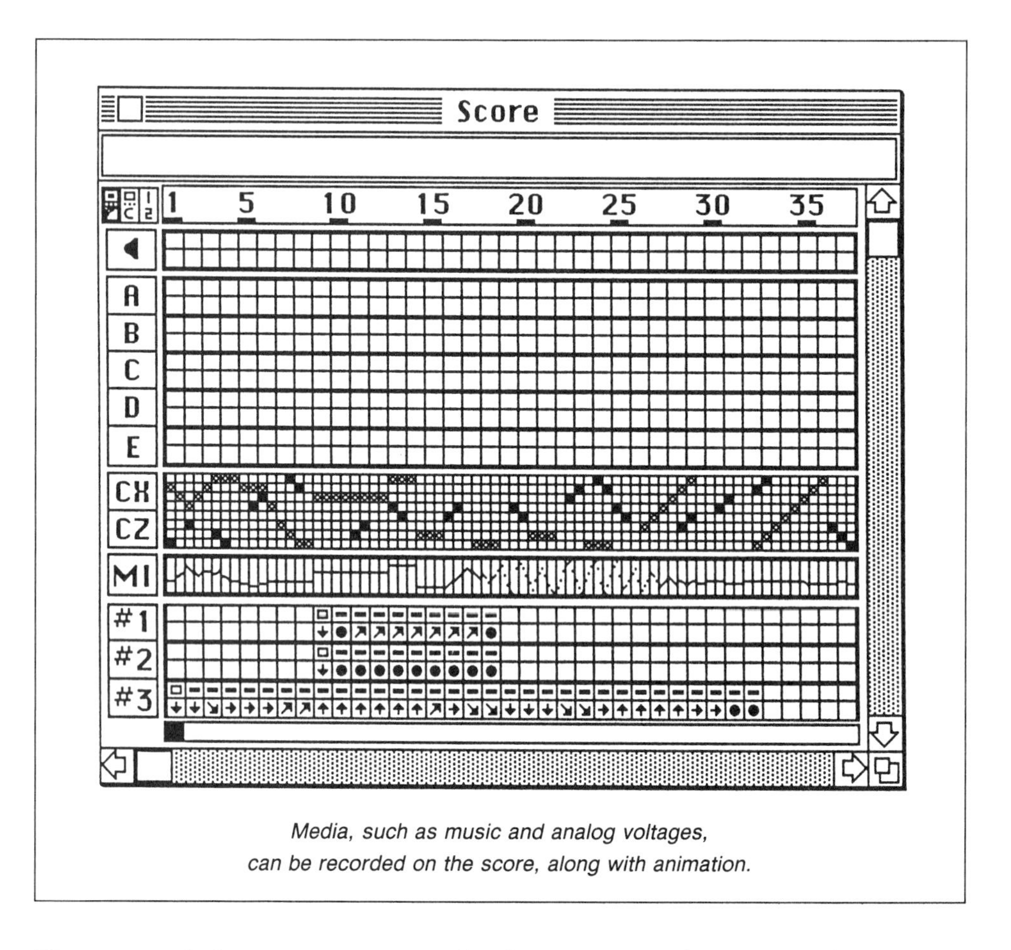

Media, such as music and analog voltages, can be recorded on the score, along with animation.

Fig. 7-16. An example of what the Zardoz score will look like.

These sort of phenomena are very hard to notate with VideoWorks (or even Zardoz), especially if you want to simulate spontaneous, natural events, but Hookup will take care of the problem and act as a general mechanism for synchronizing sound and animation.

IN CLOSING

What do sampled sounds, animation on top of still backgrounds, multiple channels of audio, random access, hi-res graphics, and hot text with animations have in common? What can you do with large buffers for audio and graphics, with VLSI support for moving pixels, with photographic-quality graphics and three other, entirely independent graphics planes, and with real-time interfacing between a user and the machine? Well, just imagine . . . .

We're driving through a city. We're suddenly attacked by stock footage of Mothra, but Godzilla comes to our rescue. We then find ourselves in San Francisco during the 1906 earthquake, and we have to get out of there as fast as possible. We awake in a hospital bed with doctors standing over us. We leave the hospital, consult our overview map, and cruise right back out onto the streets. . . .

Or, let's say we're at a World's Fair and we go into the Israeli pavilion. We find a machine that wants to tell us about how Columbus discovered America. As we go through the story, the machine confronts us with a question: Was Columbus a Jew? An entire scenario is based around the answer we give. If we answer "yes," the story of the Marranos unfolds and the importance of Queen Isabella's chief financier (a Marrano Jew) becomes evident. If we answer "no," we are referred to the Italian version of the story which, of course, is running over at the Italian pavilion. Luckily, the systems are networked together, so right from our seat in the Israeli pavilion, we can learn the Italian version of Columbus's national origins. Or, maybe we find ourselves learning the Swedish version of the discovery of America, which has Lief Ericsson landing in America a lot earlier than Columbus.

So you can see, a lot can happen with CD-I. We need to step beyond repair training and medical-school applications, and use our imaginations. Can't wait? Well, neither can we.

8

Interactive Information Technologies and Their Potential in Education

Gabriel D. Ofiesh, Ed.D.

Gabriel D. Ofiesh *is Emeritus Professor of Educational Technology at Howard University, Washington, D.C. His experience ranges from the U.S. Air Force Academy, where he was Professor of Psychology and Director of Leadership Studies, to the Air Training Command, where he pioneered the instructional systems approach to training. For the past eight years, he has served as a consultant to the Office of Foreign Relations, U.S. Department of Labor, where he is tracking new training technologies. He also has served as chief consultant in educational technology to Price Waterhouse and other multinational organizations, and has conducted training programs in instructional and new educational technologies for trainers in Panama, Costa Rica, Brazil, El Salvador, Nicaragua, Chile, Turkey, and Saudi Arabia.*

Dr. Ofiesh holds an M.S. degree in psychology from Columbia University and an Ed.D. from the University of Denver.

> It is something to be able to paint a particular picture, or to carve a statue, or to make a few objects beautiful; but it is far more glorious to carve and paint the very atmosphere and medium through which we look. . . . To affect the quality of the day—that is the highest of arts.
> *Henry Thoreau*

Western society is shifting from an industrial-based society to an information-based one. The information flood has already started, with the very volume of information creating a new barrier to its absorption. The acquisition of knowledge, which seems to progress arithmetically, is lagging far behind the geometrically expanding knowledge base. Scientific advances are thus limited because researchers are unaware of important new discoveries in fields other than their own.

As Vannevar Bush so aptly stated over forty years ago in the July 1945 issue of the *Atlantic Monthly*:

> There is a growing mountain of research. But there is increased evidence that we are being bogged down today as specialization extends. The investigator is staggered by the findings and conclusions of thousands of other workers—conclusions which he cannot find time to grasp, much less to remember, as they appear. Yet, specialization becomes increasingly necessary for progress, and the effort to bridge between the disciplines is correspondingly superficial.

The new interactive technologies, powered by the microcomputer, meet the perennial need for better ways to expand human capabilities and enhance human reasoning. They also meet the need to convert data into useful information and to transform information into *knowable* knowledge.

The microcomputer in combination with videodisc, CD-ROM, or CD-I—and who knows what next?—records, stores, and processes bits of information in milliseconds; it would take human intelligence hundreds or thousands or even millions more moments of time to accomplish the same tasks. But, more than that, the microcomputer allows us to interact with the knowledge base in ways that we could never have imagined in the past.

The microcomputer and its new interactive peripherals are the great facilitators of creativity. They have given wings to the creator by providing a pseudo-intelligence that complements the human mind. They make possible the aggregation, analysis, and processing of information necessary to creative endeavor.

The usefulness of these tools will depend on the degree of our familiarization with them. We will need to become comfortable with them and make them ours before we can put them to work for us. This takes time, and it is an involving, as well as evolving, process. First will come the tools, then the basic skill and knowledge of their use, and then, through daily exposure, proficiency will develop slowly but deliberately. In brief, we need to *expand our experience* of the tools. Someone has said that experience is not what happens to you but *what you do with what happens to you.* After we have gained experience with these tools, we will be able to intelligently analyze our needs in light of their technological potential. Then, and only then, can creativity take flight.

THE POTENTIAL IN EDUCATION

> The new technologies give us the opportunity, if we are up to it, to shape the art of education in the future. *Lester Thurow*

Thanks to the microcomputer and its peripherals, it is now possible to create, at a reasonable price, a highly interactive instructional system with characteristics never before available at any price. A new way of organizing and reorganizing the knowledge base is evolving that actually facilitates learning. This has been brought about by the very essence of the interactivity that takes place between the tools (the technology) and their users.

Throughout most of civilized history, technology has been linked with education and learning, and one can fully appreciate the interactivity of the new information technologies when they are applied to education and training. The primary concern of educators must always remain that of teaching students how to learn; and, if they can dramatize the education process, then the new technologies will ultimately have an impact on life itself.

In the past, when students did not learn and the available learning materials were not easily understood, we rarely if ever blamed the teacher or the textbook writer. The students were more often than not considered to blame. Either they were too stupid or too dense to understand "simple" explanations, or they were improperly trained and educated by previous teachers or previous schools. Educators have often declined to accept responsibility for this failure.

The new interactive technologies may well give rise to an entirely new concept of academic responsibility. The burden of responsibility for the student's learning will rest firmly on the instructional design inherent in the software and on the delivery system used to present it, rather than on the students or their former teachers. If students do not learn, something is wrong with the design. It must be thrown out and a new one tried. Or it must be "debugged" so that it will teach effectively and efficiently. Or the material must be resequenced. Or something else must be done to the program until it finally does the job. In fact, the product should be changed until it can teach its subject to practically everybody everywhere, even students that other teachers and schools have given up on.

Gerald Zacharias, who was, thirty years ago, science advisor to several United States presidents, pointed out that the basic task of educational research and development is to find out how to take the skills of great teachers, which are experienced by only a fortunate few, and disseminate them to the millions of students who are not so fortunate. As tools in the hands of committed teachers, instructors, and trainers, the new technologies provide the opportunity to produce and manage materials that will ensure learning in as many people as possible. Poor teaching and dull, boring instructors who have hidden behind routine and drill have led to much student failure and to the subsequent dehumanization of these students. The new interactive information technologies, rather than dehumanizing, will humanize. CD-I and other technologies involving multimedia information easily accessible by students offer learning in the best Socratic and tutorial tradition of teaching. Through the proper management of these tools, we can convince difficult, "stupid," unmotivated, or gifted-though-neglected students that the burden of learning is no longer solely on their shoulders but is now equally (if not more so) on the shoulders of those who consider themselves professionals in the field of education (which may someday become the science of pedagogy).

Through the new interactive technologies, we can truly individualize education. Much of what has gone under the name of individualized education has really been individual education; that is, students proceeding at their own rate but using the same interactive exercises. We are now able to create interactive exercises that adapt themselves to *different learners*. In fact, the new technologies will allow students to select their own cluster of interactive exercises—those that are most productive for them and with which they feel most comfortable. They will even be able to select the media they feel are most conducive to their own learning. In this and many other ways, individualized education will take on a completely different form than it has today. Through the new technologies, we can create truly *responsive learning environments* (to borrow a phrase from Omar Khayyam Moore) so that rather than students adapting to their learning environment, as they are currently doing, even if they are presently using interactive technology, they will be in control of the environment and the tools embedded in it. Both the environment and the tools will be responsive to the students' idiosyncratic needs and cognitive style. Not only will individualized education take on a new dramatic form, but it can finally become what we have always meant it to be.

HIGH-FIDELITY INTERACTIVITY

We know that even primitive and sterile interactivity helps learning. In numerous experiments in programmed learning, even poorly written and dull programs proved beneficial to learning because of the frequent interactivity. Research on learning behavior has found that students exposed to the computer for drill and practice, programming, and other uses feel more positive about themselves and school than those without similar access. The main advantage that the *new* micro-powered interactive technologies, which are "hot" media, have over the previous "cold" (more passive) media delivery systems, such as radio, television, and audio tapes, is their high degree of interactivity. Video, for example, can no longer afford to be a passive viewing of analog data. In fact, even analog data in the form of movies will increasingly become "cold" and will ultimately disappear as it is replaced by digital information and other media that will make visual information more readily available, manipulable, and interactive.

Already, some social changes have resulted from increased interactivity of the delivery media. The new technologies seem to facilitate a unique peer interaction, placing students in a more cooperative, collaborative environment as they work to solve problems posed by the micro-based system. The potential in the new technologies is the opportunity to change the way, nature, and scope of interactivity.

We are on the road to achieving high-fidelity interactivity, but the road has been somewhat rocky, as a review of recent history will show. The research on the teaching-learning process really started the first wave in the education and training revolution. During the last two or three decades, pioneers like B. F.

Skinner and Sidney Presser worked with the *process* of programmed instruction and defined how learning takes place in individuals. The products of their work emerged under various labels, such as instructional design and development, criterion-referenced instruction, performance-based instruction, and competency-based individualized and mediated learning (CBIM). All of these systems were flexible and adaptable, self-paced, competency-based, and, most important, *interactive*. But theirs was a sterile and primitive form of interactivity. The media then available limited the interactivity and the extent of students' involvement in the learning process.

The second wave of development was ushered in by mainframe and miniframe computers, which made possible computer-assisted instruction (CAI). To a limited degree, CAI increased the level of interactivity in the learning process. But the costs were high and the larger computers were too cumbersome to use with other media—films, sound/slides, and video recorders. Much of the CAI that has been used in schools, beyond drill practice and limited tutorial activity, has really been computer-assisted and computer-managed instruction. This is not to decry the use of individualized or structured methodology with either first or second wave interactive delivery systems. They have been moderately successful and certainly provide an alternative to conventional methodologies.

The third wave of the teaching-learning revolution had its beginnings with the microprocessor, the computer-assisted interactive videodisc, voice-recognition and voice-synthesis systems, local area networks, robot teaching machines, satellite communications, and other high-tech developments. Today the personal computer is available not only to schools, but also to people who pursue learning through nonformal education programs. As the cost of high technology has declined, its reliability and accessibility has increased, and the impact on education has the potential for being truly monumental if we are up to making it so. Instructional designers and educators from both formal and nonformal disciplines were in a particularly advantageous position to seize the opportunity—to borrow Alvin Toffler's metaphor—to become a part of the third wave. But before many of us were able to make full use of second and third wave technologies, developments in laser technology were already launching us into the fourth wave.

With such laser technologies as CD-ROM, CD-I, Smart (Laser) Cards, the optical computer, and many others that are difficult for us to foresee at present, the fourth wave of micro-based information systems is just beginning. Reasonably priced innovations and peripherals will continue to flood the market. Many of them (for example, *Dragon's Lair*) were first developed for entertainment purposes but have subsequently been exploited for their educational value. Their main attraction is their interactivity, which allows them to be individualized to meet a specific student's needs and interests.

The new technology confronts us with a new interactivity that will become increasingly transparent—students will interact with these systems without being aware that they are learning and that their behavior is being modified in the process. Yet the rapidity of these technological developments is astounding. Adapting to the telephone, radio, and television was a slow process—we could afford to take our time. The technological environment is continually changing, and, as the rate of change accelerates, it becomes increasingly more volatile. There is little time to adapt to one technology before another is there to take its place. I suggest that "interactivity"—today's buzzword—will eventually disappear, since all media will become interactive.

EMPHASIS ON FEEDBACK

In order to instruct, we have to communicate effectively. We have to know how effective our communication has been before we know how to modify our efforts to communicate. Modification must be based on knowing where, how, and why we have failed in the first effort to communicate. In order to find this out, we have to listen to feedback from students. Until the feedback loop is closed, instruction has not taken place. Very few students learn from teachers. They acquire most of their knowledge from reading, studying, analyzing, and arguing. Teachers may inspire, occasionally motivate, and even illuminate an abstract or abstruse principle, but—it bears repeating—they rarely teach anyone anything. Except, that is, in one specific kind of student-teacher relationship: the tutorial.

Now, what has all this to do with new laser technologies like CD-I? These provide delivery systems that can present teaching materials that have been developed, or programmed as we now say, in a special way. Moreover, it is the manner in which the information that goes into the machine is prepared, the size of the increments of information, the information's sequence of presentation, and its ability to branch to associated ideas, as well as how students respond to the material, that determine how effective the machine really is. We call this material "the program," and, because it is critical to the end result, the program is the heart of any CD-ROM– and CD-I–based technology.

This new technology leads students one step at a time along the learning path. They actively respond to a curriculum that has been so logically sequenced that each response takes them a little closer to the desired learning goal. Students are required to respond to questions, solve problems, or complete exercises; they then receive immediate feedback about whether their responses are correct and why. If they respond incorrectly, they are referred to additional information that allows them to correct their answers. It is this aspect of the interactive tools that comes the closest to approximating the tutorial relationship.

Because of these radical new tools, we have come a long way from the primitive notions of interaction espoused by behaviorists. Contrast the new program format with that proposed by Skinner, who argued that the ideal interactive program should be constructed so that students would make no errors in their responses. According to Skinner, information should be presented to students in very small steps and students would then be cued or prompted in such a manner that they could not help but make the correct response.

Narrow concentration on the specific features of a newly developed technique, to the exclusion of its underlying concepts, can focus attention on the tool rather than on the problems it is meant to solve. A few years ago, one problem with the process of programmed learning was that we pursued the mechanics of the teaching machine with little attention to the programmed learning process. The new technologies are not in this kind of immediate danger, though an imbalance in focus may occur in a few years if research and development studies of their potential use do not keep pace with the development of the mechanics of the tools.

Admittedly, the mechanical tools of first and second wave technologies did limit us in our ability to exploit creatively the potential of programmed instruction. I recall one of Skinner's cohorts, an expert practitioner, asserting that no instructional frame should be more than twenty-five words in length without providing students with an exercise and feedback! But Skinner and others were constrained by the tools available at that time, such as the textbook, the workbook, the multiple-choice-based teaching machine, and the mainframe computer. I am convinced that had he and his colleagues had access to these new delivery systems, they would not have restricted the concept of the constructed response to filling in a blank, drawing a diagram, solving a problem, writing a word, and the like, which determined the extent to which students could participate actively in the learning process. Now the nature of the constructed response can change dramatically, creating a new architectural form that is much more comprehensive and creative than just filling in a blank or selecting a multiple-choice response. After all, students do not live in a multiple-choice world. They are surrounded by stimuli from all parts of the cultural compass, and the new technologies allow them to interact fully with the multisensory world in which they live. But the bottom line is that all constructs—overt or covert—need the feedback that Skinner talked about, so that learning can proceed towards a predetermined destination. Without feedback, learning becomes anarchy. Therefore, we must build into the tools the feedback that allows the students to know how well they are progressing in the acquisition of knowledge and skills.

NEW DEVELOPMENTS—NEW OPPORTUNITIES

The videodisc was developed just eight years ago. The audio compact disc (CD) had its beginnings only four years ago; yet it is now considered the most successful consumer electronic product in history, and there are not enough manufacturing plants to meet the worldwide demand for CD players. Just two years ago (1984), the write-once optical disc became available; last year (1985) saw release of spin-offs from the compact disc like CD-ROM; and now CD-I is on the horizon. So within a few years, an electronics advancement designed primarily for entertainment has become a major medium for storing information and, therefore, knowledge. My guess is that you will soon be able to erase and rerecord on an optical disc, if not next year then certainly within another few years—certainly no later than the early 1990s, but probably much sooner. And within a few years, the new laser-powered computer technology will revolutionize the industry.

Bell Laboratories is now working on technologies that will enable humans to converse with computers simply by talking to them; speech synthesizers will be used to sound out anything that is typed into a computer. Already, computers are able to scan written material and convert it into simulated speech—a particularly promising development for the visually impaired. As technology makes conversation with computers a reality, educators can begin to apply interactive voice recognition and simulation to specific educational problems and situations. Can you imagine an interactive conversation with a voice recognition embedded in a CD-I disc?

So, a technological development less than a decade old will dramatically alter the peripherals tied to the microcomputer. I am convinced that CD-ROM and CD-I will become peripherals for nearly every microcomputer in the world, as soon as optical discs can be erased and rerecorded in a way similar to the way videotapes are handled by the VCR. As these new interactive delivery systems become the heart of the information revolution, they will inevitably have an impact on every form of information and education. Once erasing and rerecording capabilities are available, radical movements in information use will take place. Everyone will begin experimenting with the use of optical discs for information storage, and they will create with them new forms of interactivity.

To take advantage of these advances, educational technologists face a real challenge: to develop improved curricula that utilize the interactive capability of new technologies. Lawrence Grayson, a scientist with the National Institute of Education, suggests that ignorance about the process of instruction, rather than ignorance of technology or the high cost of technology, may become the limiting factor in the development and use of these new technologies. Microcomputer-based interactive media are valuable delivery systems, but they must be in the hands of able learning theorists who can probe the instructional process so that realistic learning materials can emerge. In the future, more resources should be committed to basic psychological research on the teaching-learning process, particularly as it applies to instructional theory, and to performing evaluative studies of programs as they are developed. There is an urgent need for educators and curriculum developers to develop sophistication in the problems and issues surrounding the emerging technologies and their impact on educational design. For educational opportunities to become increasingly available to all, it is incumbent on educational leadership to have a working knowledge of these issues as they affect institutions of learning.

It is my thesis that until pedagogy becomes scientifically based, it will not become a profession in the full meaning of the word. We know very little about how learning takes place. B. F. Skinner suggests that if we applied only 10% of the 10% we do know, we could revolutionize the educational process, but so far we have not really had a science of learning that has given us much that we can use for human resource development, and there are no formulas to apply our knowledge to new technologies.

New uses for microcomputers and videodiscs are multiplying rapidly. Simulation of long or complex processes on the computer screen—such as an assembly operation at a plant or driver safety education—can be a vital aid to students. Telephone installers are now partially trained by microcomputers that use computer graphics to simulate installation, and a micro-powered videodisc is used to instruct pilots in landing small planes—at less cost than traditional simulators for the same training. Simulation is the kind of learning activity that provides a *synergistic form of continuous interaction*, but this synergistic interactivity has been difficult to achieve without using very expensive machines—until now. A micro-powered videodisc can speed up or slow down the real time involved in a process, and it allows learning to be individualized. Now we can even simulate, and emulate, a philosopher's mental meanderings, the thought processes of a great scientist, and who knows what else? The minds of poets? Or of great composers perhaps?

We are also witnessing the proliferation of other technologies that challenge us at the chess table or test the logic of our thought processes, and these can be useful in furthering interactivity. Examples are all-electronic cameras, 8-mm VCRs, two-way cable television, fully computerized TV receivers (capable of being preprogrammed weeks ahead of time), image capturing devices, conference TVs that use picture phones, telemail, viewdata, and robots that perform repetitive or hazardous tasks.

For example, interactive videodiscs are being used to teach the skills involved in cardiopulmonary resuscitation. The interactive capabilities are activated by a light pen used to answer questions on a video screen. A mannequin wired to a microcomputer relays information to a doctor/coach who can then correct mistakes. David Hon, the designer of this interactive training device, refers to the mannequin as a "mutated keyboard." Can we then conceive of an automobile engine being wired so that it can act as a mutated keyboard for an automobile mechanic? In fact, the mutated-keyboard concept allows us to look for interactivity in all parts of our environment and thus serves to bring us closer to the world in which we live.

With such diverse and sophisticated technologies available, students can be exposed to a full range of media, from traditional graphics to video. Most important, these media can be used to develop interactive, hands-on, individualized learning. They can help trainees involve all their faculties to build and test their competencies. They can provide varied opportunities for students to use, generalize, or specialize their newly acquired skills.

An example of an innovative application is the computerization of the *Domesday Book,* the famed chronicle of medieval England commissioned in 1086, twenty years after the Battle of Hastings, by William the Conqueror. The anniversary of the *Domesday Book* marks 900 years of continuity in English government and society. The computer is being used to research the way of life portrayed in these early demographic surveys and records. Dr. John Palmer, a senior history lecturer at the University of Hull in England, has spent five years feeding the *Domesday Book* into a computer to produce computer-readable CD-ROM text. Working with him from the United States are two medieval researchers who are using a computer to reconstruct detailed maps and scholarly indices that they beam to London via satellite. The result is a far more complex and integrated portrait of Norman England than anyone could have believed possible. Imagine how the power of the researcher and student/layperson is increased by being able to "search and manipulate" all the information available on the Domesday CD-ROM disc.

Another appropriate educational example comes to us from the field of interactive telecommunications. Kathy Ferralli, a teacher, and Anthony Ferralli, an electronics engineer, are training teachers using a distance learning technology called *interactive video communications.* Their latest development permits instructors and students to communicate via a modem regardless of differences in time, location, culture, or language. By combining modems, videodisc players, a universally recognized color graphics system, and a device for the remote control of videodisc equipment, teachers are able to tutor one-on-one over long distances through fast and efficient graphics generation. The teacher manipulates graphics from the student's videodisc player using a remote control device, and the whole procedure can be saved for future student reference.

Finally, we find ourselves at the point where the new technologies demand that we think in new ways, not only about how we learn and how we use the knowledge we have, but about how the knowledge we are going to see, hear, and read will be presented to us. The parameters of old technologies, such as the printed book, forced us to think vertically and incrementally. Through the interactive videodisc, CD-ROM, and other optical mass storage systems, the presentation of knowledge will change in such a way that it will *emulate the way we think*, using associations that group knowledge segments together, rather than requiring that we adapt the way we think and read to its format. Reciprocally, the new technologies will inspire us to alter the way we think—much as microcomputers have inspired new ways of thought for the 12- and 13-year-old hackers who have challenged our theories of cognition.

We have come a long way from the sterile interactivity of the programmed text, the casual question-answer of the classroom, and the cocktail-party type of conversation. The nature of interactivity that is now available to us will be increasingly fruitful to us tomorrow when, through artificial intelligence, it will enable us to interact in "true dialogues" with "experts." We didn't dare imagine we could do this in the past, and we hesitate to imagine it even today. In the not too distant future—say in five to ten years—we will be able to conceptualize any sensory-stimulus configuration and deliver it to any student at any age level in any part of the world.

The creative tools, now available, will allow us to create innovations with such a sense of cautious urgency that the undreamt ideas of today will easily become the way we did things yesterday.

Appendices

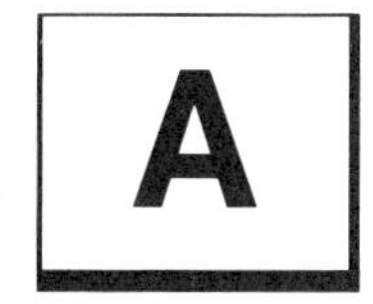

Glossary

Access To retrieve information from a storage medium. *See also* **Random access.**

Access time The time it takes to locate, retrieve, and display data after a command has been issued. This time period usually represents the worst-case situation, such as the time it takes a videodisc to search from the first to the last frame. *See also* **Random access.**

Active display The contents of the screen being displayed, as opposed to the screen contents being stored in memory for possible later display.

A/D (Analog to Digital) The conversion of information from analog format to digital format.

Address An alphanumeric or numeric label that identifies each unique location on a disc (e.g., frame numbers on videodiscs).

Address code The specific address of a particular piece of information.

ADPCM (Adaptive Delta Pulse Code Modulation) A CD-I technique used for encoding audio information.

Algorithm A precisely defined structure or sequence of actions that provides a solution to a problem.

Analog A variable that can have any value across a continuous range. Physical variables typically expressed in analog values are heat, light, voltage, and current.

Applications program (or software) A computer program designed to perform a specific task, such as accounting or word processing, as opposed to a general-purpose program, such as a computer language.

Artificial intelligence The ability of a computer to "learn" from past experience and, therefore, theoretically develop better methods of performing its assigned tasks.

ASCII (American Standard Code for Information Interchange) An internationally standardized character code system that uses an eight-bit code to identify computer keyboard characters. Seven bits (128 different characters) describe all the symbols (letters, numbers, punctuation, and control keys) used on a computer keyboard; the eighth bit may be dropped, used for parity, or used to identify nonstandard characters.

Aspect ratio The measurement of a viewing area in terms of relative height and width. Film (35 mm) is 2 × 3, television is 3 × 4, and motion pictures are 3 × 5 to 3 × 7. Variations in aspect ratio cause problems in converting between media because either something has to be cut off or there is a band of empty space.

Assemble To produce a machine-code version of a program by converting instructions written in another computer language, such as BASIC or C.

Assembly language A low-level language that uses some sort of mnemonic code to represent the machine-language instructions for a particular computer.

Audio mute The ability to suppress the audio portion of the output.

Audio track A separately addressable section of a compact audio disc, normally carrying a self-contained piece of music.

Authoring Developing the elements of an interactive program.

Authoring language A computer programming language designed around the needs of authors of interactive media.

Authoring system An aid to interactive design consisting essentially of an authoring language with a user interface that guides the author through the process of creation.

Auto stop A preprogrammed instruction that automatically stops a player when it reaches a certain point on the disc.

Automatic playback A technique by which preprogrammed control instructions are automatically executed by a player's internal computer, with the viewer responding at certain points.

Bandwidth The frequency spectrum within which signals can pass through a piece of hardware without significant attenuation. This is a limiting factor in transmitting high-speed audio and video over telephone lines and between pieces of hardware.

Base case The minimum equipment required to play a CD-I, as defined in the *Green Book*, the CD-I specification published by Philips.

BASIC (Beginner's All-purpose Symbolic Instruction Code) A high-level programming language that uses English-like statements to control a computer.

Baud The unit for expressing the relative rate at which binary data is communicated. The actual rate depends on the method of transmission.

BER (Bit Error Rate) A unit of measurement representing the number of writing errors in a given length, area, or volume of a particular storage medium.

Bit A contraction of the term *binary digit*. A bit, represented by either a 0 or a 1, is the smallest unit of information a computer deals with. Seven or eight bits combined in a certain order make up one ASCII character. *See* **ASCII.**

Bit density The number of bits that can be recorded per unit length, area, or volume of a particular storage medium.

Bit error rate *See* **BER**.

Bit map A type of computer graphic in which the largest element that can be worked with is an individual bit. A bit-mapped graphic is usually stored by representing each pixel on the screen as a bit (a 0 or a 1) in the computer's memory, as opposed to vector graphics, in which lines and other shapes exist as single elements.

Bit stream transmission A method of transmitting characters one after another in a stream. Characters are sent at fixed time intervals, with no start or stop bits.

Blanking Turning off the video signal. This is often done while a new image is being located or loaded.

Block In CD-ROM and CD-I, the user data portion of a sector.

Branch To jump from one place in a program to another, or from one video segment to another.

Branching point A decision point in a program.

Buffer An area of the computer that holds data temporarily.

Byte Eight bits.

C language A high-level computer language developed by Bell Laboratories. C produces fast, compact code that is easily moved from one type of computer to another.

CAI (Computer-Assisted or Computer-Aided Instruction) The use of a computer and an interactive program for purposes of instruction.

CAL (Computer-Aided Learning) *See* **CAI**.

Capacitance In videodisc, one of two incompatible formats. *See* **CED format**.

Capacitance electronic disc *See* **CED format**.

Cathode ray tube *See* **CRT**.

CAV (Constant Angular Velocity) A mode of videodisc playback in which the disc rotates at a constant speed, regardless of the position of the read head. *See also* **CLV**.

CBT (Computer-Based Training) *See* **CAI**.

CD (Compact Disc) A 4.72-inch (12-cm) optical disc that stores audio in a digital format.

CD-I (Compact Disc–Interactive) A multimedia system for simultaneous and interactive presentation of video, audio, text, and data. The specifications for CD-I were developed jointly by Philips and Sony.

CD-ROM (Compact Disc–Read Only Memory) A 4.72-inch (12-cm) optical memory storage medium used primarily to store computer data. A CD-ROM holds about 600 megabytes of data.

CD-ROM drive or player The device that retrieves data from a CD-ROM. It differs from a standard compact audio disc player in that it supports a higher level of error correction.

CD-RTOS (Compact Disc–Real-Time Operating System) The CD-I operating system based on OS-9, an operating system developed by Microware.

CED (Capacitive Electronic Disc) format A videodisc format developed by RCA, but abandoned in 1984.

Central processing unit *See* **CPU**.

Chapter One independent, self-contained segment of an interactive video program.

Chapter stop A code embedded in some videodiscs to signal the break between chapters, allowing specific chapters to be accessed.

Check disc The initial disc, usually of lower quality than a production disc, that is produced to confirm the accuracy of interactive program coding.

Chip, audio In CD-I, a dedicated integrated circuit designed to fulfill specific audio functions.

Chip, video In CD-I, a dedicated integrated circuit designed to fulfill specific video functions.

CLUT (Color LookUp Table) A means of compressing the amount of information required to store color pictures by allowing only a specific number of colors. The colors are stored in a table, and the color of an individual pixel in the picture is recorded as a value in the table. When the picture is reproduced, the value is looked up and its color is substituted.

CLV (Constant Linear Velocity) An alternative to CAV videodisc format. With CLV videodiscs, the speed at which the disc rotates varies from 1800 rpm to 600 rpm, depending on the distance of the read head from the center of the disc.

Color lookup table *See* **CLUT**.

Color map A table storing the definitions of the red, green, and blue (RGB) components of colors that are to be displayed on a computer monitor. *See also* **CLUT**.

Compact disc *See* **CD**.

Compact disc–digital audio standard The accepted standard, throughout the audio industry, to which all digital audio compact discs and players conform. Sometimes referred to as the *Red Book*.

Compatible Used to describe devices that can share software without modification. For example, an IBM-compatible computer is one produced by a company other than IBM but capable of running the same software that an IBM PC runs.

Composite video The various elements required to produce color video, combined into one signal.

Compressed audio *See* **Still-frame audio**.

Compression The reduction of the amount of data required to store information. There are various compression schemes that are applied to audio, video, and computer data.

Computer code Computer programs written in machine code—the instruction set used by a particular computer. Because this code operates faster and uses less memory than higher-level languages, it is used extensively in real-time applications.

Computer graphics Visual images generated by a computer.

Concealment In compact disc, the hiding of errors by an interpolation scheme. Concealment is possible only with audio or video data, where some loss can be tolerated.

Conditional branching An instruction that provides for the performance of a test, and then specifies the next instruction to be executed depending on the result of the test.

Constant angular velocity *See* **CAV**.

Constant linear velocity *See* **CLV**.

Continuous branching In videodiscs, the ability of the user to modify the presentation at any point, rather than only at specific branch points. Modification is accomplished through combining digital data and video on the same disc.

Controller A specialized computer or processor used to control the flow of data between a computer and one or more memory devices. With many personal computers (the IBM PC, for example), the controller is in the form of a board that plugs into the PC bus and into which the hard disk drive, floppy disk drive, CD player, or other memory device is plugged.

Courseware Instructional software.

CPU (Central Processing Unit) Usually a single integrated circuit (for example, a Z80, 8088, 80286, or 68000) that performs the calculations and other data manipulations in a computer.

Cross talk Generally, an unwanted transfer of energy from one circuit to another. In optical media, cross talk is the tendency of the read head to pick up information from the track adjacent to the one on which it is focused.

CRT (Cathode Ray Tube) The "screen" in a computer monitor.

Cue A pulse entered onto one of the lines of the vertical interval that results in frame numbers, picture codes, chapter codes, closed captions, and white flags on the master tape or videodisc.

Cursor An indicator on a computer's screen that marks the position at which information (for example, data typed by the user) will appear.

Cursor plane In multiplane video presentations, the plane in which the computer's cursor is presented.

Cut and paste The process of snipping a piece of text or a graphic from one location (cutting) and inserting it into another (pasting). In computers, this process is done electronically, rather than with scissors and glue.

D/A (Digital to Analog) The conversion of a digital signal to analog values.

Data Information or facts that are in a form usable by a computer.

Data capture A process by which information is electronically or electromechanically read into a computer. Use of this process avoids the drudgery of entering information by hand.

Data dump Typically, the high-speed output of all the information available on a specific subject.

Data rate The speed at which data is transferred.

Database An organized collection of information that can be accessed by a computer.

Decision point The moment in a program when the user must make a choice.

Decoder A device that selects one or more outputs depending on the combination of inputs.

Dedicated The term used to describe a piece of hardware or a program that is designed to perform a specific task. For example, a dedicated word processor is a computer that can be used only for word processing, as opposed to a general-purpose computer that can be used for word processing, among other things.

Delivery system The video and computer equipment actually used to reproduce the information on a disc.

Digital dump A process typical of Level II videodiscs, in which a digital program is stored on the disc and loaded (dumped) into the computer's memory, in order to run it.

Direct absolute RGB A picture coding scheme used in CD-I for high-quality graphics. Images are encoded on disc as red, green, and blue components using five bits for each color, plus one overlay or control bit.

Disc An optical storage medium.

DiscoVision A joint venture of MCA and IBM that introduced the videodisc in 1972.

Disk A magnetic storage medium.

Dissolve The smooth transition from one image to another.

DMA (Direct Memory Access) An input/output system in which data is transferred directly between the computer's memory and other storage devices (hard disk, CD-ROM, and so on) without going through the control unit. This results in a faster transfer.

Download The process of moving information from one computer to another. Download implies that the information is going from a "big" computer to a smaller one, and upload implies the opposite.

DRAW (Direct Read After Write) Optical discs that can be locally recorded but not erased.

Dual-channel audio The ability to play two channels, either separately or together.

Dump *See* **Data dump**.

DYUV In CD-I, a high-efficiency coding scheme used for natural pictures.

EDAC (Error Detection And Correction) An encoding technique that detects and corrects bit errors in digital data.

EIDS (Electronic Information Delivery System) The combination of hardware required to play interactive instructional material produced by the Department of Defense.

Electronic publishing The paperless delivery of information via computer.

Encode To convert information to machine- or computer-readable form.

EPROM (Erasable Programmable Read-Only Memory) A type of PROM that can be erased by exposing it to ultraviolet light.

Error correction A technique used to identify and correct errors in the storage and transfer of information.

Executable code A set of instructions, or a computer program, that can be run directly, rather than through the use of another program. Examples are any .COM or .EXE files on an IBM PC.

Fading The gradual reduction of a video or audio signal to zero (fade-out) or the gradual increase from zero (fade-in).

Feedback Information concerning the result of a function or activity. In interactive instruction, the reinforcement of correct answers or the correction of incorrect answers.

Field Half (every other scan line) of a complete television scanning cycle. Two interlaced fields make one video frame.

Formatted Data organized in a specific way to meet an established standard.

Frame A single complete picture in a video or film recording.

Frame address The numeric label identifying a particular frame. Each frame in videotape and videodisc has an address so that it can be individually located and displayed.

Frame buffer A memory device used to store the video-screen image.

Frame flutter A visual disturbance in a videodisc image that is caused when the fields from two different images are combined during editing into one frame. Flutter can also happen when a motion sequence is frozen on a particular frame of the sequence.

Frame rate The speed at which frames are scanned—30 frames a second for NTSC, 25 frames a second for PAL/SECAM.

Frame storer A device that stores one complete video frame.

Freeze frame A single frame from a segment of motion video that is held motionless on the screen.

Generic videodisc A videodisc containing a collection of information but not designed for a specific course of instruction.

Graphic overlay *See* **Overlay**.

Graphics input device A device, such as a scanner or a digitizer, that forms an image of a real object and converts it to a format that can be loaded into a computer.

Graphics output device A device, such as a computer screen or a printer, that displays the graphic images stored in a computer.

Green Book The Compact Disc–Interactive (CD-I) Standards document developed by Philips and Sony that specifies the standards for the disc format and hardware used with CD-I.

Hardware The physical electronic elements in a computer system.

Header field In CD-ROM, a 4-byte section of a data sector containing the absolute sector address (3 bytes) and the mode byte.

Hex Short for hexadecimal.

Hexadecimal A base-16 number system that would seem perfectly natural if we had eight fingers on each hand.

High-level language A computer programming language such as BASIC or C in which a command is represented by an English-like statement that translates into complex sets of assembly or machine code.

High Sierra Group An ad hoc standards group set up to establish nominal standards in the field of CD-ROM. Named after the hotel in Lake Tahoe where the group first met in the summer of 1985.

Horizontal blanking interval *See* **Horizontal retrace period**.

Horizontal retrace period The time period during which the horizontal line scan of a television screen returns to the beginning of the next line. The electron beam of the CRT is blanked out during this period, which is also known as the horizontal blanking interval.

Icon A graphic representation of a function or a task.

Image enhancement The process by which the contrast and/or detail of a video image are electronically improved.

Image planes The overlaying planes or levels upon which images can be presented. CD-I has five overlaying planes.

Insert module A piece of hardware or software (e.g., external interface connectors, videotext decoders) which, when inserted into a piece of equipment or a system, enables it to perform additional functions.

Instant jump The ability of some videodisc players to branch almost instantly to another frame located within 250 frames in either direction of the current location. Instant jumps occur without the visual disruption on the screen that normally occurs when videodisc players search to segments more than one or two frames away.

Integrated products Products that have a combination of functions integrated within a single piece of hardware (e.g., CD-I players, a computer with a built-in display).

Intelligent player A compact disc player with additional built-in computing capabilities, enabling the player to interact with the user (e.g., CD-I players).

Interactive A computer-controlled application in which the user affects and/or determines the program's flow and content by responding in some manner to the program at various predetermined points.

Interleaving A method of storing separate information sequences that allows for smooth switching between alternate courses of action. If there are three possible action sequences, for example, the first frame or field of each sequence would be laid down (1A, 2A, 3A) before the second frame or field were laid down (1B, 2B, 3B), etc. At any point the program can switch to the same spot in the alternate sequence by performing an instant jump of one or two frames.

IVD (Interactive VideoDisc) A laser videodisc program that is under computer control. *See* **Interactive**.

Keypad An input device consisting of a small keyboard containing keys that can be used to respond to the possible choices in the program.

Kiosk A stand-alone display containing a computer and/or a videodisc player, input devices, and a display screen. This type of display is often used in a retail environment for point-of-purchase, point-of-sale, and information systems.

Laser The beam of light used to write and read the information on optical media. The word *laser* stands for Light Amplification by Stimulation of Emission of Radiation.

Laser disc Commonly used to refer to a variety of reflective optical videodiscs. LaserDisc™ is a trademark for the reflective optical videodisc products made by Pioneer Electronic Corporation.

Laserfilm The trade name for the nonstandard transmissive videodisc system produced by McDonnell Douglas Electronics Company (MDEC). The Laserfilm system was designed as an on-site record system, and features a built-in compressed-audio capability.

LaserVision The trade name for the reflective optical videodisc format promoted by Philips, Pioneer, Hitachi, and others. LaserVision discs have become the industry standard.

Linear video Noninteractive video—a video designed to be played from beginning to end, without branching. Film is an example of linear video.

Machine language The lowest level of computer programming language; it is made up of a set of codes that can be read directly by the computer.

Magnetic storage Hard disk, floppy disk, or tape, upon which information is stored as variations in magnetic polarity.

Main channel On CDs, the section of the data stream that carries information, as opposed to control and display or other information which is carried in the subcode channels.

Manual operation An action that must be carried out by the system operator, apart from standard interaction.

Master The original disc, used to produce copies for distribution.

Mastering The process of producing copies from a master disc.

Matching exercise An instructional technique that requires items in one group to be paired with items in another group.

MDEC (McDonnell Douglas Electronics Company) The manufacturers of the Laserfilm system. *See* **Laserfilm**.

Menu A list of the available options from which the user can make a selection.

Menu-driven A method of computer-control in which the user selects options from a series of menus.

Mosaic graphics In CD-I, low-resolution graphics produced by repeating pixels and lines by a certain factor.

NTSC video format The color television format used in the United States and prepared by the National Television Systems Committee of the Electronics Industries Association (EIA). Commonly used in referring to broadcast-quality television.

OEM (Original Equipment Manufacturer) A term indicating equipment that is purchased from the manufacturer for use in fabricating a system for resale.

OMDR (Optical Memory Disc Recorders) A line of recorders made by Matsushita Panasonic.

Optical media Any medium that stores and reads information through the use of a laser beam.

OS-9 The real-time operating system that forms the basis for the CD-I operating system.

Overlay Computer-generated text or graphics superimposed on full-motion video.

PAL (Phase Alternation Line) video format The color television format used in all European countries except France.

Peripheral Equipment, such as a mouse, keypad, or videodisc player, that serves as an input or output device for a computer.

Picture stop The ability of certain videodisc systems to stop the player on a specific frame. Also the instruction that causes the stop.

Pit The microscopic indentation burned into the surface of an optical disc by a laser beam. Information is stored on the disc by varying the pattern of the pits.

Pixel A picture element; the smallest dot that can be produced on a video screen. The number of pixels that can be displayed across and down the screen is used as a measure of the screen's resolution.

POP (Point-Of-Purchase) and **POS (Point-Of-Sale)** Interactive video systems set up in public places to demonstrate products or encourage sales.

Postproduction The editing, encoding, authoring, and other steps necessary to prepare video information for placement on optical media.

Premastering The final stage in the preparation of a disc, when the master tape is prepared for transfer to the master disc.

RAM (Random Access Memory) That portion of the computer's internal memory in which the user can store and later retrieve information.

Random access A method of storing information so that the computer can jump directly to any point in the information without having to read the information between the current location and the target location.

Read/write capability The ability to both store information on a medium and retrieve it.

Real time A term used to describe a computer program that appears to produce the end result as soon as the inputs are received, as opposed to a program that stores the inputs for later processing and display.

Red Book The Compact Disc–Digital Audio (CD-DA) Standards document produced by Philips and Sony.

Resynthesis parameters The technical method used to regenerate audio information stored in a compressed or encoded method on storage media such as compact disc.

RGB (Red-Green-Blue) A type of color output to a computer display, consisting of separate signals (on separate wires) for the red, green, and blue components of the picture. An alternative to *composite* video.

ROM (Read Only Memory) That portion of the computer's memory (usually a relatively small portion) that stores information the user can access but not change.

RS-232C A standard for the serial interface between a computer and its peripherals.

RTOS (Real-Time Operating System) The CD-I operating system developed by Microware.

Run length coding An encoding technique that compresses the data required to store a given screen of information.

Scan A method of rapidly browsing through video by displaying only a portion of the information that the read head passes over.

SCSI (Small Computer Standard Interface) A standard for interfacing various peripheral devices to a computer.

Search To move directly to a specific frame on a disc.

Search time The time required to locate a specific piece of information on a disc.

SECAM (Sequential Couleur ã Memoire) video format The color television format used in France and Russia based on the sequential recording of primary colors in alternate lines.

Sector In CD-ROM, a block or frame complete with synchronization and header field.

SelectaVision® The trade name for RCA's CED videodisc format.

Sequencing The order of presentation of information in an instructional program.

Simulation The realistic portrayal of a situation or function, for the purpose of familiarizing people with a situation without them having to actually experience it first.

SMPTE time code An 80-bit standardized edit time code adopted by the Society of Motion Picture and Television Engineers.

Speech synthesizer A device that converts text or other computer input to speech.

Spiderweb A learning exercise in which branching from different decision points may lead the user to some of the same results. The flowchart thus resembles a spiderweb.

Sprites Movable shapes (usually relatively small) on the computer screen that are defined by the program and under its control.

Stand-alone systems Equipment that is capable of functioning on its own. Level II videodisc and CD-I players are typical stand-alone systems.

Step frame The ability of a videodisc player to move forward or backward one frame at a time while in the freeze-frame mode of operation.

Still-frame Information recorded on a frame or track of a videodisc and played repeatedly in order to display it as a single motionless image.

Still-frame audio A method of storing several seconds of voice-grade audio in a single frame on a disc by digitally encoding it.

Submenu In a hierarchical menu system, a menu one or more layers down from the main menu. A submenu allows the user to branch to new information without returning to the main menu.

Subroutine A self-contained program segment to which the program branches in order to perform a specific task, after which control is returned to the main program.

Super TOC (Table Of Contents) In CD-I, the information in the first track on the disc. Synonymous with disc label. The super TOC lists the disc type and format, the status of the disc (single or part of an album), the data size, and the position of the file directory and bootstrap, which are required to start up the CD-I player.

Table of contents *See* **TOC** and **Super TOC**.

Time code An eight-digit addressing system for videotape frames that is recorded on the spare track or is inserted in the blanking interval.

Timed video still A frame of information left on the screen for a specific period of time.

TOC (Table Of Contents) Subcode information in the lead-in area of the compact disc that identifies the number of tracks and indexes, as well as their timing and duration.

Touch screen A computer screen that functions as an input device by dividing the screen into an x-y coordinate system and by reading the location that the user touches.

Transmissive optical videodisc format A transparent videodisc originally developed by Thompson/CSF that allows the laser beam to pass through the disc to the detector. The only transmissive system currently on the market is the McDonnell Douglas Electronics Company's Laserfilm system. *See* **Laserfilm**.

Treatment The form or manner of presenting a subject in film or video. There are five major kinds of treatment: didactic, documentary, dramatic, graphics, and combination. Also, a narrative description of a proposed video program.

Tree A learning exercise structured so that one decision leads to more choices, which in turn branch to yet more choices. Thus, the flowchart resembles a tree.

User interface The manner by which the user relates to the computer or to the program running on it. The interface may include a hardware device such as a touch screen or mouse, or software concepts such as menus and icons.

Vertical blanking interval Lines 1 through 21 of video field 1 and lines 263 through 284 of video field 2, in which frame numbers, picture stops, chapter stops, white flags, closed captions, and so on may be encoded. These lines do not appear on the screen but are used to maintain image stability and to enhance image access.

Vertical interval time code *See* **VITC**.

VHD (Video High Density) A grooveless capacitance videodisc format that uses a broad stylus to pick up data.

Videodisc format One of four types of videodisc: reflective optical videodisc, transmissive optical videodisc, capacitance electronic disc (CED), and video high density (VHD).

VITC (Vertical Interval Time Code) The SMPTE time code inserted in the vertical interval between the two fields of a tape frame. This method eliminates errors that occur from tape stretch when using longitudinal time code.

WORM (Write Once Read Many) A type of permanent optical storage that allows the user to record information on a disc and read it, but not erase or change it.

Write-once media *See* **WORM**.

Yellow Book The Compact Disc–Read Only Memory (CD-ROM) Standards document developed by Philips and Sony.

B

Resources

LISTING OF RESOURCES BY CATEGORY

Applications Designers

Active Learning Systems
Alamo Learning Systems
American Interactive Media (AIM)
ByVideo
Comsell, Inc.
Deltak Training
Discovery Systems
Earth View, Inc.
Haukom Associates
Health EduTech
Industrial Training Corporation
Info Disc Corporation
Info Express Inc.
Interactive Television Company
Interactive Training Systems (ITS)
IVID Communications
IXION, Inc.
KnowledgeSet Corporation
Learning Resource Network
Malone & Tessman Productions
Maritz, Inc.
Pacific Interactive
Perceptronics
The Record Group (TRG)
Rodesch and Associates
Ropiequet & Associates
Sallis & Associates
Society for Visual Education (SVE)
Systems Impact
WICAT Systems

Consulting Services

Haukom Associates
Info Express, Inc.
IXION, Inc.
Kent, Rick
Ofiesh, Gabriel D.
Sallis & Associates
Michael Uretz

Data Preparation

Digital Audio Disc Laboratories
Info Express Inc.
KnowledgeSet Corporation
VideoTools

Discs Mentioned in This Book

Army Tank Gunnery Training (Perceptronics)
Astronomy Disc (Optical Data Corporation)
BioSci Disc (VideoDiscovery)
Business Disc (MITEC)

Cardio Pulmonary Resuscitation (CPR) System (American Heart Association, Actronics)
Citizen Kane (Criterion Collection, The Voyager Company)
Cityguide Discs (Technology For People)
Decision Point (Digital Equipment Corporation)
Delaware Music Series (University of Delaware)
Developing Your Financial Strategies (Starship Industries, International Education Corporation)
Discursions (MIT Demonstration Disc, Architecture Machine Group at MIT)
Dragon's Lair (Cinematronics)
EPCOT Center Videodiscs (WED Enterprises)
Firefox (ATARI)
Great Whales/Sharks (National Geographic)
How to Watch Pro Football (Optical Programming Associates)
Kidisc, The First National (Optical Programming Associates)
King Kong (Criterion Collection, The Voyager Company)
KnowledgeDisc (KnowledgeSet Corporation)
MACH 3 (Mylstar Electronics, Inc.)
Motourist Discs (Technology For People)
MysteryDisc, "Murder Anyone?" (Vidmax)
National Air and Space Museum Discs (Smithsonian, Visual Database Systems)
National Gallery of Art Disc (Videodisc Publishing)
Paris on Videodisc (ADIC Presse)
Pearlstein Draws the Artist's Model (Interactive Media Corporation)
Quarterhorse (Electro-Sport, Rodesch and Associates)
Space Ace (Cinematronics)
Space Archives, Space Disc series (Optical Data Corporation)
Thayer's Quest (RDI Video Systems)
Versatile Organization (Wilson Learning)
Video Encyclopedia of the 20th Century (CEL Communications)
Video Jukebox (Cable Video Jukebox, Inc.)
Vincent van Gogh (NA Philips)
Welding Simulator (IXION, Inc., Academy of Aeronautics)

Disc Pressing

3M (videodiscs, DRAW discs, compact discs, CD-ROM, erasable data discs)
Crawford Communications, Inc. (DRAW discs)
Digital Audio Disc Corporation (compact discs)
Discovery Systems (videodiscs, DRAW discs, compact discs, CD-ROM)
LaserVideo, Inc. (videodiscs, compact discs, CD-ROM)
Pioneer (videodiscs, CD-ROM)
Publisher's Data Service Corporation (CD-ROM)
Sony (videodiscs)
Spectra Image (DRAW discs)
Technidisc (videodiscs, DRAW discs, compact discs, CD-ROM)

Disc Publishers and Distributors

Note: While the following companies do originate some discs themselves, each also has an extensive catalog of discs that it distributes. All of the discs mentioned in this book that are available for purchase can be found at one or all of these companies. Write or call for a catalog. (See the alphabetical listing for addresses and phone numbers.)

Pioneer Artists Society

VideoDiscovery
Voyager Company

Hardware

Acorn Company
Allen Communication, Inc.
Amiga
Apple Computer, Inc.
ATARI, Inc.
AT&T
BCD Associates, Inc.
ByVideo
Commodore Electronics, Ltd.
Digital Equipment Corporation (DEC)
Digital Research, Inc.
Digital Techniques
Drexler Technology Corporation (SmartCards)
Droidworks
EECO, Inc. (still-frame audio devices)
Hallock Systems Company
Hewlett-Packard
Hitachi Sales Corporation of America
IEV Corporation
Interactive Training Systems (ITS)
International Business Machines, Inc. (IBM)
IXION, Inc.
JVC
LaserData (still-frame audio devices)
Laserfilm Systems (LASERFILM videodiscs and MDEC videodisc systems)
Learning International, Inc.
McDonnell Douglas Electronics (LASERFILM videodiscs and MDEC videodisc systems)
MicroTrends, Inc.
Mindset
NCR Corporation
NEC America, Inc.
NEC Corporation
New Media Graphics
Nissei Sangyo America, Inc.
Online Computer Systems, Inc.
Optical Data Corporation
Optical Disc and Video Systems Company (OMDR distributor)
Optical Disc Corporation (DRAW disc systems)
Panasonic Industrial Company (OMDR videodisc systems)
Philips Subsystems and Peripherals, Inc.
Pioneer Electronic Corporation
RDI Video Systems
Signetics
Stride Micro (formerly Sage)
Sun Microsystems, Inc.
Video Associates Labs
VideoTools
Visage, Inc.
Visual Database Systems
Whitney Educational Services

Premastering Services

Image Premastering Services, Ltd. (transfers single-frame images to videotape)

Publications and Conferences

Disc Topics (periodical)
Educational Technology (periodical)
E-ITV (periodical)
IICS Journal (periodical)
Institute for Graphic Communication (conferences)
Mini-Microsystems Magazine (periodical)
Nebraska Videodisc Group (conferences, newsletter, workshops)
Optical Information Systems (bimonthly periodical)
Optical Information Systems Update (biweekly periodical)
Optical Memory News (periodical, conferences)
Society of Applied Learning Technology (conferences)
Video Computing (periodical)

Video Manager (periodical)
Videodisc Monitor (periodical, special reports)

Research and Development

Architecture Machine Group, MIT (now called The Media Laboratory)
Bell Telephone Laboratories

Software

Apple Computer, Inc.
Ashton-Tate
Computer Access Corporation
Control Data Corporation (PLATO)
Electronic Information Systems (EIS)
Hewlett-Packard
Info Express Inc.
Interactive Television Company
Interactive Training Systems (ITS)
IXION, Inc.
MacroMind
Microsoft Corporation
Microware Systems, Inc.
New Media Graphics
Online Computer Systems, Inc.
Optical Data Corporation
OWL International, Inc.
VideoTools
Whitney Educational Services

ALPHABETIC LISTING OF RESOURCES

3M Optical Recording Project
3M Center 223-5S
St. Paul, MN 55144-1000
612/733-6334
(Disc pressing)

Academy of Aeronautics
La Guardia Airport
New York, NY 11371
718/429-6600
(*Discs)

Acorn Company
23 East 10th Street, Suite 414
New York, NY 10033
212/673-3333
(Hardware)

Active Learning Systems
7801 East Bush Lake Road, Suite 350
Minneapolis, MN 55435
612/831-6916
(Applications)

Activenture Corporation (*see* KnowledgeSet Corporation)

Actronics
810 River Avenue
Pittsburgh, PA 15212
412/231-6200
(*Discs)

ADIC Presse
Paris, France
(*Discs)

AIM (*see* American Interactive Media)

Alamo Learning Systems
1850 Mount Diablo, Suite 500
Walnut Creek, CA 94596
415/930-8520
(Applications)

Allen Communication, Inc.
140 Lakeside Plaza II
5225 Wiley Post Way
Salt Lake City, UT 84116
801/537-7800
(Hardware)

American Heart Association
7320 Greenville Avenue
Dallas, TX 75231
214/750-5300
(*Discs)

American Interactive Media (AIM)
11111 Santa Monica Boulevard,
Suite 700
Los Angeles, CA 90025
213/473-4136
(Applications)

Amiga (*see* Commodore)

Apple Computer, Inc.
20525 Mariani Avenue
Cupertino, CA 95014
408/996-1010
(Hardware, software)

Architecture Machine Group
(New name: The Media Laboratory)
Massachusetts Institute of
Technology
20 Ames Street, E15-331
Cambridge, MA 02139
617/253-0338
(*Discs, research and development)

Ashton-Tate
20101 Hamilton Avenue
Torrance, CA 90502-1319
213/329-8000
(Software)

ATARI, Inc.
PO Box 61657
Sunnyvale, CA 94088
408/745-2000
(*Discs, hardware)

AT&T Bell Laboratories
Consumer Products
2002 Wellesley Boulevard
Indianapolis, IN 46219
317/352-6124
(Hardware)

BCD Associates, Inc.
7510 North Broadway Ext'n.,
Suite 205
Oklahoma City, OK 73116
405/843-4574
(Hardware)

Bell Telephone Laboratories
Interactive Video Systems Research
Department
600 Mountain Avenue,
Room 3D-474
Murray Hill, NJ 07974
201/582-7833
(Research and development)

ByVideo
225 Humboldt Court
Sunnyvale, CA 94089
408/747-1101
(Applications, hardware)

Cable Video Jukebox, Inc.
3550 Biscayne Boulevard, Suite 711
Miami, FL 33137
305/573-6122
(*Discs)

CEL Communications
515 Madison Avenue
New York, NY 10022
212/421-4030
(*Discs)

Cinematronics
1841 Friendship Drive
El Cajon, CA 92020
619/562-7000
(*Discs)

Commodore Electronics, Ltd.
1200 Wilson Drive
West Chester, PA 19380
215/431-9100
(Hardware)

Computer Access Corporation
26 Brighton Street
Belmont, MA 02178
617/484-2412
(Software)

Comsell, Inc.
500 Tech Parkway

Atlanta, GA 30313
404/872-2500
(Applications)

Control Data Corporation
8100 34th Avenue South
Bloomington, MN 55440
612/853-5022
(Software)

Crawford Communications, Inc.
506 Plasters Avenue
Atlanta, GA 30324
404/876-8722
(Disc pressing)

Deltak Training
1751 Diehl Road
Naperville, IL 60566
312/369-3000
(Applications)

Digital Audio Disc Corporation
1800 North Fruitridge Avenue
Terre Haute, IN 47804
812/466-6821
(Disc pressing)

Digital Audio Disc Laboratories
1860 Heather Court
Beverly Hills, CA 90210
213/276-5334
(Data preparation)

Digital Equipment Corporation
(DEC)
305 Foster Street
Littleton, MA 01460
617/486-2111
(Hardware)

Digital Equipment Corporation
(DEC)
Educational Services
12 Crosby Drive
Bedford, MA 01730
617/276-1367
(*Discs)

Digital Research, Inc.
60 Garden Court
Box DRI
Monterey, CA 93942
408/649-3896
(Hardware)

Digital Techniques
10 "B" Street
Burlington, MA 01803
617/273-3495
(Hardware)

Disc Topics
Pioneer Video, Inc.
Industrial Sales Division
5150 East Pacific Coast Highway,
Suite 300
Long Beach, CA 90804
213/498-0300
(Publications, conferences)

Discovery Systems
555 Metro Place North, Suite 325
Dublin, OH 43017
614/761-2000
(Applications, disc pressing)

Drexler Technology Corporation
2557 Charleston Road
Mountainview, CA 94043
415/969-7277
(Hardware)

Droidworks
PO Box CS 8180
San Raphael, CA 94912
415/485-5000
(Hardware)

Earth View, Inc.
6514 18th Avenue Northeast
Seattle, WA 98115
206/527-3168
(Applications)

Educational Technology
Publications
140 Sylvan Avenue

Englewood Cliffs, NJ 07632
201/871-4007
(Publications, conferences)

EduTech (*see* Health EduTech)

EECO, Inc.
1601 East Chestnut Avenue
PO Box 659
Santa Ana, CA 92702-0659
714/835-6000
(Hardware)

E-ITV (Educational and Industrial Television)
CS Tepfer Publishing Company
51 Sugar Hollow Road
Danbury, CT 06810
203/743-2120
(Publications, conferences)

Electronic Information Systems (EIS)
6925 South Union Park Center, Suite 465
Midvale, UT 84047
801/561-1800
(Software)

Electro-Sport
3170 Airway Avenue
Costa Mesa, CA 92626
714/979-1553
(*Discs)

EPCOT Center (*see* WED Enterprises)

Future Systems, Inc. (*see* Videodisc Monitor)

Hallock Systems Company
267 North Main Street
Herkimer, NY 13350
315/866-7125
(Hardware)

Haukom Associates
2120 Steiner Street
San Francisco, CA 94115
415/922-0214
(Applications, consulting)

Health EduTech
7801 East Bush Lake Road
Minneapolis, MN 55435
612/831-0445
(Applications)

Hewlett-Packard
1819 Page Mill Road
Palo Alto, CA 94304
415/857-2139
(Hardware, software)

Hitachi Sales Corporation of America
OEM Division
1200 Wall Street West
Lyndhurst, NJ 07071
201/935-5300
(Hardware)

IBM
InfoWindow Marketing
PO Box 2150
Atlanta, GA 30055
404/238-4250
(Hardware, software)

ICS-INTEXT (*see* Deltak Training)

IEV Corporation
254 West 400 South, Suite 280
Salt Lake City, UT 84101
801/531-0757
(Hardware)

IICS Journal
2120 Steiner Street
San Francisco, CA 94115
415/922-0214
(Publications, conferences)

Image Premastering Services, Ltd.
1408 Northland Drive, Suite 401

Mendota Heights, MN 55120
612/454-9622
(Premastering services)

Industrial Training Corporation
PO Box 6009
Rockville, MD 20850
800/638-3757
(Applications)

Info Disc Corporation
Four Professional Drive, Suite 134
Gaithersburg, MD 20879
301/948-2300
(Applications)

Info Express Inc.
6141 Northeast Bothell Way,
Suite 103
Seattle, WA 98155
206/486-8028
(Applications, consulting, data
preparation)

Institute for Graphic
Communication (IGC)
375 Commonwealth Avenue
Boston, MA 02115
617/267-9425
(Publications, conferences)

Interactive Media Corporation
165 West 46th Street, Suite 710
New York, NY 10036
212/382-0313
(*Discs)

Interactive Television Company
1800 North Kent Street, Suite 920
Arlington, VA 22209
703/525-5625
(Applications, software)

Interactive Training Systems (ITS)
4 Cambridge Center
Cambridge, MA 02142
617/497-6100
(Applications, hardware, software)

International Education Corporation
No longer in business
(*Discs)

International Interactive
Communications Society (*see* IICS
Journal)

IVID Communications
7030 Convoy Court
San Diego, CA 92111
619/576-0611
(Applications)

IXION, Inc.
1335 North Northlake Way
Seattle, WA 98103
206/547-8801
(Applications, consulting, hardware,
software)

JVC
4042 Black Fin Avenue
Irvine, CA 92714
614/660-9294
(Hardware)

Kent, Rick
15515 77th Avenue West
Edmonds, WA 98020
206/745-5400
(Consulting)

KnowledgeSet Corporation
2511 Garden Road, Building C
Monterey, CA 93940
408/375-2638
(Applications, data preparation,
*discs)

LaserData
10 Technology Drive
Lowell, MA 01851
617/937-5900
(Hardware)

Laserfilm Systems
McDonnell Douglas Electronics
Company
PO Box 426

Street Charles, Missouri 63302
314/925-4116
(Hardware)

LaserVideo, Inc.
1120 Cosby Way
Anaheim, CA 92806
714/630-6700
(Disc pressing)

Learning International, Inc.
PO Box 10211
Stamford, CT 06904
203/965-8400
(Hardware)

Learning Resource Network
PO Box 3416
Durham, NC 27702
919/683-8050
(Applications)

MacroMind
1028 West Wolfram
Chicago, IL 60657
312/327-5821
(Software)

Malone & Tessman Productions
2127 44th Avenue
San Francisco, CA 94116
415/664-0497
(Applications)

Maritz, Inc.
1375 North Highway Drive
Fenton, MO 63099
314/225-4000
(Applications)

McDonnell Douglas Electronics Corporation (*see* Laserfilm Systems)

The Media Laboratory (*see* Architecture Machine Group)

Microsoft Corporation
16011 Northeast 36th Street
PO Box 97017
Redmond, WA 98073-9717
206/882-8080
(Software)

MicroTrends, Inc.
650 Woodfield Drive, Suite 730
Schaumburg, IL 60173
312/310-8852
(Hardware)

Microware Systems, Inc.
1866 Northwest 114th Street
Des Moines, IA 50322
515/224-1929
(Software)

Mindset
965 West Maud Avenue
Sunnyvale, CA 94085
408/737-8555
(Hardware)

Mini-Microsystems Magazine
2041 Business Center Drive, Suite 109
Irvine, CA 92715
714/851-9422
(Publications, conferences)

MITV (Maryland Instructional Television)
11767 Bonita Avenue
Owings Mills, MD 21117
301/337-4210
(*Discs)

Mylstar Electronics, Inc.
No longer in business
(*Discs)

NA Philips (*see* Philips Subsystems and Peripherals, Inc.)

National Education Training Corporation (*see* Deltak Training)

National Geographic Society
Educational Services
Department 86

Washington, DC 20036
800/368-2728
(*Discs)

NCR Corporation
1700 South Patterson Boulevard
Dayton, OH 45479
513/445-5000
(Hardware)

Nebraska Videodisc Group
PO Box 83111
Lincoln, NE 68501
402/472-3611
(Publications, conferences)

NEC America, Inc.
8 Old Sod Farm Road
Melville, NY 11747
516/753-7000
(Hardware)

NEC Corporation
1414 Massachusetts Avenue
Foxboro, MA 01719
617/264-8438
(Hardware)

NETC (*see* Deltak Training)

New Media Graphics
279 Cambridge Street
Burlington, MA 01803
617/ 272-8844
(Hardware, software)

Nissei Sangyo America, Inc.
1701 Golf Road
Rolling Meadows, IL 60008
312/981-8989
(Hardware)

Ofiesh, Gabriel D.
4031 27th Road North
Arlington, VA 22207
703/243-4271
(Consulting)

Online Computer Systems, Inc.
20251 Century Boulevard
Germantown, MD 20874
800/922-9204
(Hardware, software)

Optical Data Corporation (formerly Video Vision)
66 Hanover Road
Florham Park, NJ 07932-0097
800/524-2481
(*Discs, hardware, software)

Optical Disc and Video Systems Company
1142 Manhattan Avenue, Suite CP-27
Manhattan Beach, CA 90266
213/546-4826
(Hardware)

Optical Disc Corporation
17517-H Fabrica Way
Cerritos, CA 90701
714/522-2370
(Hardware)

Optical Information Systems (Bimonthly)
Optical Information Systems Update (Biweekly)
Meckler Publishing
11 Ferry Lane West
Westport, CT 06880
203/226-6967
(Publications, conferences)

Optical Memory News
E. S. Rothchild, Publisher
PO Box 14817
San Francisco, CA 94114-0817
415/681-3700
(Publications, conferences)

Optical Programming Associates
445 Park Avenue
New York, NY 10022
212/508-2745
(*Discs)

Optical Recording Project (*see* 3M)

OWL International, Inc.
14218 Northeast 21st Street
Bellevue, WA 98007
206/747-3203
(Software)

Pacific Interactive
1010 Turquoise Street, Suite 206
San Diego, CA 92109
619/488-6300
(Applications)

Panasonic Industrial Company
Computer Products Division
Optical Disc Department
One Panasonic Way
Secaucus, NJ 07094
201/392-4602
(Hardware)

Perceptronics
6271 Variel Avenue
Woodland Hills, CA 91367
818/884-7470
(Applications, *discs)

Philips Subsystems and Peripherals, Inc.
100 East 42nd Street
New York, NY 10017
212/850-5011
(*Discs, hardware)

Pioneer Artists Society
200 West Grand Avenue
Montvale, NJ 07645
201/573-1122
(Disc publishers and distributors)

Pioneer Electronic Corporation
5000 Airport Plaza Drive
PO Box 22711
Long Beach, CA 90801-5711
213/420-5700
(Disc pressing, hardware)

Publisher's Data Service Corporation
2511 Garden Road
Building C
Monterey, CA 93940
408/372-2813
(Disc pressing)

RDI Video Systems
No longer in business
(*Discs, hardware)

The Record Group (TRG)
3300 Warner Boulevard
Burbank, CA 91510
818/953-3211
(Applications)

Rodesch and Associates
12445 Greens East Road
San Diego, CA 92128
619/485-5544
(Applications, *discs)

Ropiequet & Associates
PO Box 31273
Seattle, WA 98103
206/522-5522
(Applications)

Sallis & Associates
148 Northeast 57th Street
Seattle, WA 98105
206/522-5522
(Applications, consulting)

Signetics
811 East Arques Avenue
PO Box 3409
Sunnyvale, CA 94088-3409
408/991-2000
(Hardware)

Smithsonian Institution
National Air and Space Museum
Records Management Division
Washington, DC 20560
202/357-3133
(*Discs)

Society for Applied Learning Technology (SALT)
50 Culpeper Street
Warrenton, VA 22186
703/347-0055
(Publications, conferences)

Society for Visual Education (SVE)
1345 Deversey Parkway
Chicago, IL 60614-1299
312/525-1500
(Applications)

Sony Corporation of America
Sony Drive
Park Ridge, NJ 07656
201/930-6177
(Disc pressing)

Spectra Image
540 North Hollywood Way
Burbank, CA 91505
818/842-1111
(Disc pressing)

Starship Industries
605 Utterback Store Road
Great Falls, VA 22066
703/430-8692
(*Discs)

Stride Micro (formerly Sage)
PO Box 30016
Reno, NV 89520
702/322-6868
(Hardware)

Sun Microsystems, Inc.
3803 East Bayshore Road, Suite 175
Palo Alto, CA 94303
415/965-7800
(Hardware)

Systems Impact
4400 MacArthur Boulevard, Suite 203
Washington, DC 20007
1-800-822-INFO
(Applications)

Technidisc
2250 Meijer Drive
Troy, MI 48084-7111
Plant/Customer Service: 1-800-321-9610
Sales: 1-800-521-6714
(Disc pressing)

Technology For People (TFP)
1533 Lakeshore Drive
Columbus, OH 43212
614/488-6837
(*Discs)

University of Delaware
Office of Computer-Based Instruction
Newark, DE 19716
302/451-8161
(*Discs)

Michael Uretz
Advanced Systems, Inc.
155 East Algonquin Road
Arlington Heights, IL 60005
312/981-1500
(Consultant)

Video Associates Labs
3933 Steck Avenue, Suite B105
Austin, TX 78759
512/346-5781
(Hardware)

Video Computing
PO Box 3415
Indialantic, FL 32903
306/768-2778
(Publications, conferences)

Video Manager
Knowledge Industry Publications
701 Westchester Avenue

White Plains, NY 10604
800/248-5474
(Publications, conferences)

Video Vision (*see* Optical Data Corporation)

Videodisc Monitor
PO Box 26
Falls Church, VA 22046
703/241-1799
(Publications, conferences)

Videodisc Publishing
381 Park Avenue South, Suite 1601
New York, NY 10016
212/685-5522
(*Discs)

VideoDiscovery
PO Box 85878
Seattle, WA 98145-1878
206/547-7981
(*Discs, disc publishers and distributors)

VideoTools
445 Calle Serra
Aptos, CA 95003
408/476-5858
(Data preparation, software, hardware)

Vidmax
36 East 4th Street
Cincinatti, OH 45202
513/421-3999
(*Discs)

Visage, Inc.
12 Michigan Drive
Natick, MA 01760
617/655-1503
(Hardware)

Visual Database Systems
614 Bean Creek Road
Scott's Valley, CA 95066-3314
408/438-8396
(Hardware)

Voyager Company
2139 Manning Avenue
Los Angeles, CA 90025
800/446-2001
(*Discs, disc publishers and distributors)

WED Enterprises
1401 Flower Street
Glendale, CA 91201
818/956-6500
(*Discs)

Whitney Educational Services
415 South El Dorado Street
San Mateo, CA 94402
415/340-9822
(Hardware, software)

WICAT Systems
Training Systems Division
PO Box 539
1875 South State Street
Orem, UT 84057
801/224-6400, Ext. 209
(Applications)

Wilson Learning
2009 Pacheco Street
Santa Fe, NM 87505
505/471-6500
(*Discs)

Xerox Learning Systems (*see* Learning International, Inc.)

**Discs* indicates that specific discs by this company are mentioned in the book.

Index

D

E

F

W

Y

Z